# Electronic Emotion

# Interdisciplinary Communication Studies

Series Editor: Professor Colin B. Grant

# Volume 3

**PETER LANG**

Oxford • Bern • Berlin • Bruxelles • Frankfurt am Main • New York • Wien

Jane Vincent &
Leopoldina Fortunati (eds)

# Electronic Emotion

## The Mediation of Emotion
## via Information and
## Communication Technologies

PETER LANG

Oxford • Bern • Berlin • Bruxelles • Frankfurt am Main • New York • Wien

**Bibliographic information published by Die Deutsche Bibliothek**
Die Deutsche Bibliothek lists this publication in the Deutsche
Nationalbibliografie; detailed bibliographic data is available on the
Internet at http://dnb.ddb.de›.

A catalogue record for this book is available from The British Library.

Library of Congress Cataloging-in-Publication Data:

Electronic emotion : the mediation of emotion via information and
communication technologies / [edited by] Jane Vincent and Leopoldina
Fortunati.
    p. cm. -- (Interdisciplinary communication studies ; 3)
  Includes bibliographical references and index.
  ISBN 978-3-03911-866-3 (alk. paper)
  1. Information technology--Psychological aspects. 2.
Telematics--Psychological aspects. 3. Human-computer
interaction--Psychological aspects. 4. Emotions. 5. Identity
(Psychology) I. Vincent, Jane, 1956- II. Fortunati, Leopoldina.
  T58.5.E425 2009
  302.2301'9--dc22
                        2009006247


ISSN 1661-8645
ISBN 978-3-03911-866-3




© Peter Lang AG, International Academic Publishers, Bern 2009
Hochfeldstrasse 32, CH-3012 Bern, Switzerland
info@peterlang.com, www.peterlang.com, www.peterlang.net

Printed in Germany

# Contents

# Acknowledgements

The editors wish to thank the University of Udine, its Doctoral Programme in Multimedia Communication and the Faculty of Education for sponsoring the meeting of the Pordenone Group of Scholars at which this book was conceived. This publication is supported by COST and their staff are acknowledged for their assistance together with the COST 298 Chair Bartolomeo Sapio and Grant Holder Tomaz Turk. The series editor, Colin Grant of the University of Surrey's Faculty of Arts and Human Sciences, is thanked for his advice and guidance during the production of this book.

COST – the acronym for European COoperation in the field of Scientific and Technical Research – is the oldest and widest European intergovernmental network for cooperation in research. Established by the Ministerial Conference in November 1971, COST is presently used by the scientific communities of 35 European countries to cooperate in common research projects supported by national funds. Web: www.cost.esf.org

ESF provides the COST Office through an EC contract.

COST is supported by the EU RTD Framework programme.

# Introduction

LEOPOLDINA FORTUNATI AND JANE VINCENT

## Introduction

We believe that the study of the role of emotion in the relationship that human beings have with ICTs (information and communication technologies) is of great value to many academic disciplines and a book that explores this topic of emotion and ICTs is well overdue. A book such as this can give a more complete picture of these relationships and also help to explain at least some of the ways human interaction with ICTs is currently shaped. As Adam Smith pointed out in his first book, *The Theory of Moral Sentiments* (1759), emotion should be seen as the glue that keeps together the fabric of society. Indeed, no social science might be complete without also comprehending and addressing the reasons of the heart, of which reason, according to the famous quote of Blaise Pascal, knows nothing (O'Connell, 1997). If interaction should be seen as (more or less) a blend of emotion and reason then, we assert, communication should also be considered in the same way. Words, most commonly used for interaction, are also part emotion and part reason and this means that one can talk not only of emotional intelligence as proposed by Salovey and Mayer (1990), and Goleman (1995, 1998, 2001, 2006), but also of emotional communication. Communication cannot exist without emotion and this implies again that the study of information and communication technologies cannot be complete without looking at emotion.

A large corpus of studies on the social practices of use, general behaviours, attitudes, opinions, and representations of ICTs has emerged in recent years (Haddon et al., 2005; Castells et al., 2007; Ling, 2008). At the same time, a body of literature on the nature and the role of emotion had also been established by the work of anthropologists, sociologists, neuroscientists, and cognitive and social psychologists, such as that which examined links between the technology and socioemotional uses of electronic media (Rice and Love, 1987; Short, 1976). It is probably only by putting together these many different areas of study that one can begin to understand the complex nature of emotion. Even if we are still far from a theory of emotion that encompasses this complexity, there are today several theories and approaches that can be applied to the study of emotion. In addition a growing mass of research and reflections is also available on the emotional relationship between human beings and ICTs (Fortunati, 1995a; Fortunati and Manganelli, 1998; Contarello et al., 2008; Vincent, 2005, 2006, 2007).

In exploring the evolution of the status of machines one should acknowledge that they have become 'smart' and 'intelligent' (Gershenfeld, 1999), as well as 'emotional' and 'affective' (Simon, 1967), by progressing the process of imitation and simulation of human beings. This process of anthropomorphisation to which machines are subjected is the proof of how much human beings invest symbolically and emotionally in them. As Benjamin pointed out (1936), the specific uniqueness – the aura of machines – has diminished as a result of serialisation and mass production. However, this does not prevent machines from receiving and embodying emotions and from attracting symbolic meanings and metaphors. In fact the machines are also included within the individual's aura, where they become further charged with personal and social meanings. Humans have, of course, tried since ancient times to narrate the multifarious facets of their relationship with technology and in this example of humans, emotion and ICTs the social construction and representation of machines embodies these archetypes and myths.

A specific section of this corpus of work is dedicated to the role of emotion in information and communication for as we have discussed emotion plays an important role in the process of obtaining information. According to Bateson (1972), information might be understood as the perception of a difference out of which one produces a further difference. On a physiological level, the process starts from the organs of sense and from the sensation, which might be transformed by perception through the movements of the body. On a cognitive level the information is produced in relation to other information that is on a comparative basis with other established differences. This shows the relational and the emotional nature of information and explains why information has so often embraced narration (Carey, 1988), and entertainment (infotainment) (Brants, 1998).

Emotion plays a very important role in communication too, as is shown by the large field of studies on emotional communication (Andersen and Guerrero, 1998). Human beings use a wide range of languages to communicate, not only verbal languages, but also non-verbal and paralinguistic language, including facial expressions, bodily gestures and manifestations (such as perspiration, blushing, shaking), posture, tone of voice, and so on. The expression of emotion flows through all these languages, although the non-verbal and paralinguistic language might be understood as immediate signs rather than an intentional codex of emotional states. But while non-verbal and paralinguistic channels directly convey emotion, verbal language conveys emotion indirectly since it does so in terms of its personal and social representation. However, verbal language is most usually activated because, as Fussell (2002), underlines, non-verbal cues provide insufficient information for expressing the full range of emotion. Of course there is a lot that is not overtly communicated when expressing emotion and so the emotions are also conveyed by paralinguistic and non-verbal channels. Words also play an important role in expressing the nuance or to convey the depth of emotion, although often in an indirect way to achieve this un-communicated part of the emotional experience and feelings. Words can provide precise descriptions of the emotional states and more detailed information on the specific intensity and shape of the emotions that are experienced. A limit of non-verbal and paralinguistic channels is that they convey the immediate and personal emotions, while the verbal channel helps in expressing emotions experienced in the past or which have simply been imagined. Rime et al. (2002) show that people often talk with their partner, friends and family about the emotional experiences that occurred to them in their life and past emotional experiences are a main issue in the agenda setting of self-help groups and therapeutic settings. Fussell and Moss (1998) point out that people also talk about the emotional experiences of other people they do not know personally such as personalities and celebrities. There is much heated and lively debate on the opposing problem: that of the design of machines capable of imitating emotional expressions. These affective machines are a topic of research on which it would appear important advances are frequently achieved, and indeed, this maybe perturbing, given that: '[...] the capacity for emotions', as Evans points out (2001, p. 99), '[...] is often considered to be the main difference between humans and machines'. But this is not the main point here; rather, the development of the topic of this book is to analyse the relationship between emotions and information and communication technologies.

In this introduction we look first at ICTs and separately at emotion, in order to understand more about what they are and how they are socially constructed. After focussing on these two issues and clarifying terms and approaches, we analyse the relationship between emotion and information and communication technologies. Finally we introduce the sections and the chapters that constitute this book.

## The Information and Communication Technologies

As cultural artefacts ICTs have inherited, like all the other machines, some very powerful myths, such as that of Prometheus and Icarus, and an archetypical fabric, rich in metaphors and symbols. The strong combination of imagination, myth and science has nourished over time the narrative of science fiction, whose authors have managed a very rich terrain of invention and foresight. At the same time ICTs are part of the whole history of machines that have enabled the development of industrialisation, and, as they are so closely connected to economic development they are also seen as its engine, especially with regard to globalisation. Indeed, most capital production passes through the use of machines and the more sophisticated they are, the better they seem at enabling increased production and added value to economies.

In common with other machines, ICTs have been strictly connected with the myth of progress and of modernity and with a linear vision of time. Furthermore, all over the world, machines are perceived as contributing to a 'rational' environment. This rational environment however, is far from being an automatic application of scientific principles since the development of innovations often springs from the: '[...] technician's more intimate experience of the behaviour of matter and mechanisms' (Nye, 2007, p. 10). The rationality supported by machines is also limited by the fact that those who interact with ICTs – e-actors (Fortunati et al., 2009), are in general not interested in a deep comprehension of the technological world (Fortunati et al., 2008). In fact they use tools, systems, and devices without knowing how they work, and they are not interested in knowing technology in itself (Longo, 2003). The limitation of the rationality of machines, in the case of ICT, is also strengthened by another element. Being intellective machines, ICTs attract all the content connected to communication and information, which, as we have seen above, is made up in part by rationality and in part by emotion and all the features linked to it, not least ambiguity. We discussed earlier how words are considered to be part emotion and reason, and thus ambiguity is an inevitable consequence of this emotion and dualism that is the essence of language.[1] Furthermore, the ambiguity of the language is growing because the media language is frequently hyper-codified and this increases the level of ambiguity in language at a mass level. Fortunately,

---

1    Think, for example, of the semantic halo surrounding words or of double meanings, euphemisms, lapsus, metaphors, and symbols. Language is able to clarify concepts and ideas, but it also creates ambiguity and may be unfathomable as well as serving to avoid dealing with certain areas of reality. In particular, Jung and De Laszlo (1958) help us to understand that the nature of the symbol is the expression of an underlying relationship between two different things or concepts apparently not having any connection. So the illustration of this relationship implies an emotional tension, because in a certain sense the symbol surprises the interlocutor and requires a supplementary cooperation. Metaphors too have the capacity to describe unusual or unexpected relationships between things. They are all operations that are based on intuitions and emotion and aim to provoke new ideas and at the same time emotion. This burden of the structural ambiguity of the language, due to the eruption of emotion – the geological commotions of thought, as Marcel Proust called them – is strengthened also by a series of obstacles which menace the circularity of communication. Sociological and communication studies have enumerated these obstacles: non-coincidence of codes, social tendency to narcissism, scarce awareness of the balance between originality and redundancy, limited interest by the interlocutor, taboos and the not said, power differences, defence from a possible control, blurring boundaries between public and private dimensions, and so on.

however, as more people attain higher levels of education the more they are able to deal with the process of hyper-codification of language, and in particular with its rhetorical expression and use. But it is important to note that if the ambiguity of language becomes structural to the language itself, its modalities (spoken and written) will be affected to a different degree. In principle, the written form of communication is more likely to be ambiguous because it lacks the cues which especially convey emotion and on which spoken communication usually depends. Whereas the ambiguities in oral communication might be resolved through non-verbal cues, the written form has more difficulty in doing so. In semiotic models (Eco, 1976), misunderstandings, equivocal and interpretative dissimilarities and such aberrant decoding, are considered not as pathology, but rather as a structural eventuality of the communication process. It is not surprising therefore that misunderstanding is common in communication, even if generally it is not so common as to prevent the effective exercise of communication, as Evans points out (Evans, 2001, p. 8).

This structural ambivalence of technology is one element that may explain why the history of the relationship between Western societies and technology has always been highly emotional (Longo, 2003). The two faces of machines and their strong engine of valorisation and modernization on the one hand, and the strong engine of archetypisation and mythologisation on the other, have in turn created two audiences, made up of enthusiasts and detractors (Eco, 1964). These two opposing fronts – those in favour of technology and those against – are understandable if we take the view that technology is situated in the realm of novelties. As such it arouses on the one hand the desire of the new and the different and on the other, the refusal to accept it at all as a consequence of the defence of the status quo. There is a process, explored by Freud (1920) and by Piaget and Inhelder (1966) that is about the flow of the new into the old, or the well known. This process, generally, is played in a binary way between the pole of curiosity, rarity, new risk and uncertainty on the one hand, whilst on the other it includes old habits, stability certainty, security and safety. Detractors of technologies generally fear and refuse innovation, and according to Ferrarotti (1970), in European history intellectuals often expressed hostility towards technology. They believed machines were connected to material and thus to servile labour and may be it this belief that has contributed to problems associated with the common negative attitudes of educated people towards technology. In contrast to this considered view, enthusiasts submit to new technological products, without checking them too much to find out if they represent an improvement or not, arriving at a kind of 'technological compulsiveness'. According to Mumford (Mumford, 1967, p. 12), the myth of the machine is based on the socially shared perception that the benefits of what the machine produces are greater than the human costs it inflicts. This suggests that actually power justifies any human cost; this attitude is strengthened by the mythological narration, where technologies are archetypical of an attitude towards nature, which is of the dominion over nature, and at the same time of the challenge to gods' (and women's) creative power (Fortunati, 1995b). The history of technology, we argue is a history of challenging power, of contesting authority, of overcoming nature's limits, and of human initiative and creativity. Among the various reasons at the basis of this position is the great role played by power. In Western societies, formally under masculine domination, technology has been seen as the site of power, of power over nature, power over other human beings, and of power of control and domination.

It is of no surprise, therefore, that in Western history a great deal of emotion has been invested in technology, since these societies have always focused a great deal on the imagination, design, and production of machines (Longo, 2003). Some epistemologists, such as Maldonado (2005), have observed that it took several centuries for Western societies to move away from this opposition and to be less visceral. At the same time, the structural and emotional ambivalence of technology has allowed the creation of a science fiction that is a virtual field of experimentation, imagination, and innovation. There has been more innovation and foresight capacity in the novels and stories of great science-fiction writers, such as Asimov or Le Guin, than in many of the research and development departments of industry. With regard to the machines one continues to take in this air of ambiguity that is made up of rationality and imagination. The effort of intellectualisation and rationalisation of the world that is necessary for machines to work inevitably leads to disenchantment, according to the hypothesis originally advanced by Max Weber in the essay *Science as a*

*Vocation* (1918–19), but then the archetypical and mythological fabric which surrounds them produces a re-enchantment (Griffin, 1988). It is interesting to consider this mechanism of re-enchantment from another cultural point of observation, such as the Chinese one. From the first research on this topic in China, it would appear that people seem to have a more pragmatic relationship with technology (Tarantino, 2008). The lack of emotional investment in technology by Chinese may explain why they are focused on practical issues, such as postures while using technology, costs, and so on. In contrast with the Chinese society, Western societies have instead always put a lot of emotional investment in technology. It is this torment, this emotional investment in technology that inspired Mosco's analysis on the technological sublime (Mosco, 2004).

This ambivalent attitude towards technology and the interplay between imagination and science, emotion and rationality, is particularly evident with regard to ICTs. These ICT technologies have always been exploited by the power systems because of their emotional state and they represented the ideal loci to convey social tensions and conflicts. Powerful catalysts and transformers of conflicts, tensions, desires and expectations, the information and communication technologies have often been able to serve as a means of pacification at the social level. In fact ICTs have enabled a lot of the emotional and rational life to become dislocated at a virtual level as ICTs mediated entertainment and then information, communication, education, and so on. Emotional repertoires have been worked into the narrative plots of the cinema and television as well as into the style and ways information is proffered. People have begun to consume electronic emotion through cinema and television products. The mechanism of the spatial dislocation of emotion on the cinema and TV screens has provoked in turn an emotional investment by the general public in the technologies themselves. The emotional intensity experienced in looking at films or TV films and soap operas was in many cases also directed towards the devices that were mediating them: the cinema and the television screens.

Although the emotional relationship with technology has been characterized so deeply in the history of the technological development in the West, the overall corpus of research on the social aspects of technology largely refers to the rational side of e-actors' behaviour, social practices, attitudes, and opinions. As discussed above, very few research studies have until now focused on this relationship between emotion and information and communication technologies.

## Emotion

There are several definitions, approaches and theories of emotion, especially since the study of emotions has become an inter-disciplinary field where several subjects converge. Emotions can be defined from a neurobiological point of view or, typically within psychology, in behavioural forms. Other ways to understand emotion can be the cognitive or the anthropological and sociological approaches. Each of these elaborate aspects and characteristics of emotion it would, indeed, be wise to combine. But we are still far from being able to produce a comprehensive approach to emotion, which could integrate all these different disciplines. With regard to social sciences, anthropologists (Lutz and White, 1986; Kay and Svasek, 2005) have formulated the cultural theory of emotion, according to which emotions are understood to be learned behaviours that are transmitted culturally. They assert that people belonging to different cultures are supposed to experience different emotions or emotion in different ways.

In addition, the history of sociological thought is rich in seminal reflections and analyses on emotion. From Durkheim to Simmel, from Mead to Elias, from Goffman to Wright Mills: all these authors have proposed very illuminating analyses depicting the experience, the social norms and functions, and the social meanings of emotion. To Durkheim (1912) we owe the understanding of the notion of emotions as important agents and factors of social cohesion. Emotion is transmitted, learned and internalized in salient and collective rites and ceremonies and serves to set up solidarity and morality. Simmel (1983) develops further the analysis of the relationship between emotion and social cohesion. He looks deeply into numerous emotional processes such as gratitude, fidelity, infidelity, jealousy, envy, forgiveness, trust, hate and hostility, and he distinguishes between pre-sociological emotion and sociological emotion such as fidelity and, less strongly, gratitude, which are very important in the maintenance of social cohesion. Mead (1934, 1938) argues that emotional experience is a cultural acquisition and is at the basis of the personal life and social interactions. He considers emotion as social facts, forms of action and communication that are made up of and sustained by relational processes. His perspective is that emotions are the 'emerging' part of relationships and are part of the personal experience of self. Individuals in fact enter into contact through emotion and in turn their social relations produce other emotions. Elias (1939) sees emotions as social constructions and comes back to link their role to the functioning of the social structure and stratification. According to him, a specific structure of emotional sensitivity and repertoires corresponds to each social structure. The regulation of emotions depends on their functionality in different social systems. Goffman (1956, 1961, 1969) looks at the normative component of emotion, proposing very useful notions, such as emotional congruence, affective deviance and rules of expressions. So, while Durkheim, Simmel, Mead and Elias focused the attention on the productive function of emotion in enhancing the social structure and its cohesion, Goffman was more interested in the normative process, to understand how the regulation of emotion expresses and allows social control. Drawing on his work, Hochschild (1979), and Elkman (1982) have elaborated the notions of emotional norms and of norms of exhibition or expression. Wright Mills (1951), by analysing a particular social layer, white-collar workers, reads this case like a vivid example of the commodification of emotion. He depicts how white-collar workers are obliged to sell not only their labour-force, but also their personality, since they must submit themselves to strong control and pay for these very high emotional costs.

The integration of the analyses carried out by these classical authors enables us to explore emotion in a more complete and complex way, in the light of social practices, changes, movements, structures and stratification. Although the sociology of emotion still lacks a true epidemiology and social epistemology of emotion (Thoits, 1989), recently sociologists (Denzin, 1984; Thoits, 1989, 1992; Flam, 1990; Kemper, 1990; Barbalet, 1998; Hochschild, 1979, 1983, 1989, 2002, 2005; Stets and Turner, 2006) and social psychologists (Parrots, 2000; Niedenthal et al., 2006) have tried to focus mainly on the role of social norms in building the emotional repertoire of a society. Their purpose was to better understand the social construction of the emotional life. In particular, they have tried to study how the social construction of emotion varies according to different epochs and societies, including their sub-elements, such as classes, gender, generations, and so on, and have elaborated some theories such as the social interactionism (Kemper, 1978, 1987), and the control of affection (Heise, 1979, 1986).

According to Hammond (1990) emotions are organising principles of social differentiation and stratification. Feminist theory has proposed the opposite notion: emotions are fundamental elements of the labour, that a certain tradition has called 'emotion work' and 'emotional labour' (Hochschild, 1979, 1983, 1989, 2005, 2006), and another 'immaterial labour' (Fortunati, 1981), which indicate the large section of domestic labour consisting in caring, nurturing and love. It has been possible to assume emotion as immaterial work because a theoretical analysis of the process of reproduction (domestic labour, nurturing and prostitution) in respect to and beyond the Marxist categories was elaborated. While Marx clearly saw the domestic sphere as an unproductive sphere, this feminist analysis has demonstrated that the production of immaterial goods and services is a crucial stage inside the whole process of production and reproduction. The productivity of the emotional part of immaterial labour was understood as producing and/or reproducing the commodity most precious for capital, the labour force. Emotion when considered part of immaterial labour in this way happens to be totally inside the process of value-added production. Recently, Hardt and Negri (2000, pp. 290–2), inspired in part by the feminist tradition, have taken up this concept of immaterial labour. They have recognized that, in addition to services, cultural products, knowledge, or communication, immaterial labour is constituted also by the affective labour of human contacts and interactions.

In the last twenty years, many studies have highlighted the tendency to export immaterial labour and specifically emotion work from the domestic sphere in the whole capitalist system of the most industrialized countries (Lazzarato, 1997). Immaterial labour has spread like a virus to the whole economic system, whereby the dynamics that govern the reproduction of the labour force has been exported (Hochschild, 1997). The extension of the functioning of the domestic sphere to the capitalist system has happened because this is the only way to increase the productivity of labour exploiting more deeply the new generations of labour-force in globalisation. This has implied the increasing development of the emotional sphere and experience in the society which was carried out through the extension and intensification of the working day and by making people consume television, cinema, the internet, mobile phones and computer games for the purposes of communication, information, education (e-learning), knowledge and organization. Thus, there has been a machinisation and a serialisation of emotion, which integrate the commodification framework depicted by Hochschild (1983). As Alquati (1970, p. 128), points out, machinisation embodies the secondary properties of immaterial labour, since the machine is more docile (mild) and less expensive and so in the process of technological intensification is able to guarantee growing savings in living labour. In the production of material goods, where processes like computerization, the rapid implementation of technological innovations, and the increasing importance of information have quickly taken hold, not only has immaterial labour spread, but also new mechanisms of social control and management of social representations were introduced. Immaterial labour with its prominent emotional part has become even more productive for capital and in this new social and global order, emotions, as immaterial labour, have changed in part their statute, becoming also largely mediated electronically.

This analysis enhances Denzin's (1990), point that the issue of emotion should be treated as a problem of political economy. There is in fact a true political economy of emotion which is skilfully orchestrated in many ways through socialization, consumption and especially with the aid of the mass media. Those with the economic and political power are well aware of how important the social management of emotion is both to guarantee the productive functioning of social structure and its cohesion. But there is another reason supporting Denzin's concern: that of the mass ownership and use of information and communication technologies that contribute a great deal to shaping social cohesion, class stratification, and the mechanisms of production of value-add (Fortunati, 2006). Collins (1975, 1990) helps us to understand some of the mechanisms put in motion to differentiate the access to the production and consumption of emotion among the various social strata. He points out that power and status might be recognized as the basic dimensions of social stratification. Thus, people with high power and status obtain emotional energy from their dominant position whilst those with low power status and position lose emotional energy. Traditional users and audiences of ICTs, being in a position of subjection, were structurally destined to find their emotion diminished in the sense that their emotion is 'vampirized' or bled dry. The consequence of the consumption of mediated emotion is the regimentation of the various emotions themselves. Direct (humans-to-humans) emotion risks, in practice, being annihilated in order to give the advantage to indirect humans-to-machine emotion.

## The Relationship between Emotion and ICTs

Given the premises asserted above, reconstructing the role played by emotion in the contexts, situations and sites where ICTs are used and in which society and technology co-construct themselves reciprocally, becomes strategic not only at a sociological, but also at an economic level. Furthermore, it is also strategic to reconstruct the cultural frames, ideologies, and social representations that institutionalize and formalize emotion in social life. The questions we would like to address at this point are: What does an electronic emotion concretely mean? How does emotion change when mediated by the information and communication technologies? What does emotion become when represented and treated in ICT content? How are the production and the consumption of electronic and mediated emotions articulated? How do people experiment with them? What emotional investment do people express in the ICTs?

A mediated emotion is an emotion felt, narrated or showed, which is produced or consumed, for example in a telephone or mobile phone conversation, in a film or a TV programme or in a website, in other words mediated by a computational electronic device. Electronic emotions are emotions lived, re-lived or discovered through machines. Through ICT, emotions are on one hand amplified, shaped, stereotyped, re-invented and on the other sacrificed, because they must submit themselves to the technological limits and languages of a machine. Mediated emotions are emotions which are expressed at a distance from the interlocutor or the broadcaster, and which consequently take place during the break up of the unitary process which usually provides the formation of attitudes and which consists of cognition, emotion and behaviour. If one shares Harré's and Gillett's (1994), approach that emotion words function to perform emotional acts, mediated emotions remain separated by immediate behavioural processes carried out towards and/or with the interlocutor(s). This of course represents a serious limitation of this experience. Keeping in touch emotionally with people who are distant from oneself as well as consuming narrated emotions have in a certain sense displaced emotion from the point of its production and its link to the consumption. So people as audiences might consume more emotion, but produce and transmit less emotion. This paradoxically might produce a block or a loss (in Collins' words, 1990), of emotional energies. In no other historical moment has emotional literacy been a topic so discussed and operationalized. At the same time this penetration into people's lives of virtual, electronic emotion has rarefied the emotional sphere, making it more difficult to detect. On the other hand, although suffering as a result of this limitation, mediated emotions have enhanced the cognitive sphere and in particular the imaginative sphere. They have allowed people to explore many ways to express, invent and learn emotions, by overcoming spatial and temporal limitations. In addition, they constitute a precious emotional experience that has also enriched social relationships.

In many cases mediated emotion is emotion described verbally or in writing. The boundaries of these emotions may lie inside a radio programme or in telephone and mobile conversations, as well as in email, instant messages, chats, forums, blogs, social networks, music (MP3, iPod). These mediated emotions exist orally or in writing without vision, so without non verbal cues. A mediated emotion can also be produced by broadcasters and consumed by audiences in a TV programme, in a film, or in an online or mobile video and pictures. Here there are images in motion, orality, music and also non verbal cues, but here also audiences are only able to consume emotions, expressed and formalized by others. So in addition to the features of mediated emotions, which is an important point because they shape the emotional fluxes at social level, there is another aspect that is equally important, if not even more: the control one has over the flux, shape and model of the emotion broadcast and consumed.

Information and communication technologies are in some ways the creators and diffusers of emotion, since their capabilities suggest emotional styles and practices of expression. At the same time they function as repositories of electronic and mediated emotions. This makes them very powerful instruments that are able to capture and orient emotional flows. ICTs allow broadcasters the possibility to invent and propose ideologies, models, and concepts, to control and shape attitudes and social thinking about the most adequate emotional behaviour. Consequently ICTs show, teach and impose on the general public in their products what they construe to be the right facial expressions, gestures and words, to express emotion. The everyday life use of ICTs is thus contributing to the shaping of people's emotional life and to continuously re-structuring emotional repertoires.

Within the universe of mediated emotions there are some branches of research, but the major part of this field has still to be explored. We are at the infancy of studies on emotion and ICTs. However, it is important to acknowledge that the current transformations which are occurring in the information and communication technologies and mass media – media convergence, interactivity, citizen-journalism, net TV, You Tube, social networks, blogs – are re-defining the power relationship between broadcasters and the media groups on the one hand and the audiences on the other hand. Audiences are taking initiatives in information, in music and video production and so on and the power relationship is going in part towards the general public rather than the media groups, telecommunications and operators. If one adds to the definition of social stratification another dimension that is the access to economic resources, as the mainstream sociological tradition suggests (Gallino, 1993; Scott and Marshall, 2005), this would imply that consequently the ownership of technologies also becomes an important point in the social rites of power. It is in fact on this ownership of ICTs (which includes also the access to and the use of ICTs) that e-actors are building a new power. This means it is quite possible an era is starting in which the use of ICTs by e-actors is not automatically configured as a loss, but more likely as an attainment of emotional energy.

In this book we look at emotion not only to be like the constitutive part of immaterial labour, but also as a way to access the patterns of meaning which e-actors apply in the relationship with the information and communication technologies. Why now a book on emotion and technology? Because after years and years of thinking and debating about rational action as the fundamental tenet of social behaviour, sociology has produced a new series of reflections about the necessity to understand emotions, as socially constructed and culturally constructed artefacts. As a consequence, not only has a sociology of emotion emerged as a fertile and recognized field of studies, but emotion was also discussed as a variable, influencing the methodological level as well as empirical (Holland, 2007; Bondi, 2003, 2005). These factors have made possible the articulation of a more complex model of the social actions, which includes the rational and normative aspects as well as the emotional aspects. Indeed, it has become specifically important to apply this more sophisticated model to the terrain of technology, given the complex relationship at the symbolic and cultural level towards technology that people have also expressed in the West, as discussed above.

The study of emotion in relation to information and communication technologies was even more urgent, given that ICTs are technologies that enable the conveyance of meanings, symbols, values and emotion. It is not surprising, therefore, that this theme of the relationship between emotion and ICTs is attracting the increasing attention of the scientific community within multiple disciplines. In this book we aim to explore current research questions regarding this theme examining various multi-disciplinary approaches towards a common goal of understanding the relationship between emotion and ICTs. This book addresses the issue of electronic emotion by bringing together the work of a group of international academics whom are part of a global community researching ICTs and peoples' participation in broadband society. The chapters of this book represent a very important moment of dialogue and discussion within this community that has emerged from an international seminar of leading researchers in Pordenone Italy. This seminar was unique in the degree to which it was an international and interdisciplinary event focusing on a single theme – Emotion and ICTs – and in the ways it tried to produce a true conceptual integration of different approaches and disciplines on this issue. The Pordenone Group of Scholars were probably the first multi-disciplinary group of this kind to meet to discuss emotion and ICTs and this book presents the discourse of members of this group in the form of newly written material on their respective research topics. The ICT world as explored by these academics from various disciplines including social sciences, computer and engineering sciences and design, comprises personal and social relationships, devices, signs and communication information practices and its users. The papers that are presented and discussed in this volume include current theoretical frameworks, contemporary research projects, issues in approaches to the role of emotion in the information society, ways of analysing people's emotional experience of ICTs and potential social implications. The current aim is to build upon the dialogues and networks that emerged from this seminar, crossing and integrating many different cultures as well as conceptual perspectives with a view to developing the new theoretical position on emotion and ICTs that is emerging from the work of this network.

In our introduction thus far we have appraised the research on emotion with regard to the ICTs and now we turn to the contents of this volume. In this book we seek to understand ICTs through the lens of emotion, and we explore this proposing a model implying three themes. First the emotion those ICT users feel when using or not using the different devices; secondly the emotion mediated through the contents that ICTs convey and thirdly the emotional investment that users put into the ICTs and the devices. This examination towards the understanding of emotion with regard to ICTs is made at both an inter-disciplinary and cross-cultural level. These three themes are analysed separately and outlined in the specific sections containing the individual chapters.

## Theme 1: The Emotion ICT Users Feel when Using, or Not Using, Different Devices

In this section, introduced by Leopoldina Fortunati's chapter 'Old and New Media, Old Emotion', research on emotion and ICT is discussed with reference to specific case studies from Italy, Germany and Japan. These offer powerful examples of how emotion influences and affects our everyday lives as well as discussing research methodologies for collecting and analysing the data. In addition to the mobile phone which is the foremost information and communication technology examined in this section of the book, Fortunati's paper explores landlines, fax and answering machines, the mass media (television and radio) and the computer/internet. After discussing why it is necessary to look at the emotional sphere in order to understand the complexity of the relationship between human beings and the communication and information technologies, Fortunati briefly analyses some aspects of the complex emotional relationship between individuals and machines both at a diachronic and synchronic level. She explores the results of quantitative and qualitative research studies carried out in Italy that included examination of the emotional impact of the telephone ring, the telephone voice and the answering machine. Voice, she asserts, is the primary means of telephone communication, since it is through the voice that various emotions are carried out in addition to being expressed by means of words. Fortunati also presents a cross-cultural research project on the emotion conveyed by different media and describes the map of different emotions felt by users towards fixed and mobile telephone, television, computer, internet, fax, answering machine and radio, in five European countries – France, Italy, Germany, Spain and the UK. Furthermore, she describes the main different emotional flows towards ICTs among these countries and the relationships between emotion and ICT ownership and use. In her findings she explores how users generally experience positive emotions towards ICTs whilst mobile phones and television in particular also present a negative emotional side.

It is the negative side of the mobile phone that Joachim Höflich looks at in his chapter 'Mobile Phone Calls and Emotional Stress'. Distress is a powerful emotion often present in the domestication and integration of mobile phones in everyday routines, as a consequence of the disruption of the rules of good manners regulating social behaviour in the public space. The topic of distress has been well documented (cf. Haddon, 1998; Ito et al., 2005), but it is the way in which Höflich looks at how people manage this distress that is new. In fact he uses observational studies carried out in Germany to investigate disturbing moments that occur in peoples' lives as a result of mobile phone use, what their reactions are and how these disturbances might create emotional stress. He analyses the mobile phone with regard to it being an obtrusive medium, arguing that it is usually associated with disturbing the public sphere. Theoretically, the study is based on an ethnomethodological approach that looks at everyday life strategies to produce normality in critical situations. Höflich applies breaching experiments to explore changes to the normal behaviours of mobile phone users, recording his observations of the effects of disturbances that are introduced via the mobile phone.

Satomi Sugiyama's chapter, 'Decorated Mobile Phones and Emotional Attachment for Japanese Youths', looks, instead, at the positive side of the mobile phone. She focuses on the image-making ways in which Japanese adolescents express emotion in the presentation of the self. In a certain sense, this author continues the works begun by Fortunati, Katz and Riccini (2003), Ling and Pedersen (2005), and Ito, Okabe and Matsuda (2005), but she focuses on the relations between self and identity on the one hand and technology on the other. In particular she analyses mobile communications that are among the barriers for modern mediated communication and puts the role of the self at the centre of the emotion and at the centre of the processes of the use of this technology. The paper analyses the mobile phone as a locus of mediating emotion by defining oneself in the context of the presentation of self in public places. Emotion is examined in giving material support such as mobile messages, photographs and decorations in Japan. The essence of this decorative mobile sphere is a sign of distinction in which Japanese youths also express the specific ways they possess and use their mobile phones.

## Theme 2: Emotion Mediated through the Contents ICTs Convey

Three different approaches to exploring this topic are included in this section introduced by Naomi Baron, who examines the attempts to express emotion by written markers in computer mediated communication, such as the first use of the emoticon ☺ in 1982. The section goes on to examine the ways that emotion is used to articulate interaction with ICTs, how emotion can change attitudes and be used to manage the many different emotion states of the people as they communicate. The information and communication technologies that are in the front line here are the internet and the cinema (trailer). In her chapter, 'The Myth of Impoverished Signal: Dispelling the Spoken Language Fallacy for Emoticons in Online Communication', Baron deals with the linguistic dimensions of expressing emotion in computer mediated communications. In body-to-body encounters we have multiple modalities for conveying meaning, including voice, gesture, and physical distance from our interlocutor. However, once communication becomes written only tangible graphemes are available for indicating general semantic intent and emotional nuance. Baron's paper explores the introduction of so-called emoticons, (such as smiley and frowny faces formed with punctuation marks), in the history of on-line communication. She argues that the original motivations for introducing emoticons may no longer be valid. The chapter demonstrates how changing user-ship of an information communication technology, along with domestication of that ICT, may result in people rethinking the value of linguistic contrivances such as smileys and frownies to express emotion.

Baron's article arises in the complex dynamics between co-presence and online communication, speech and written communication, while Maria Bortoluzzi's paper, 'An Inconvenient Truth: Multimodal Emotions in Identity Construction', poses the problem of the complex framework made up by multimodal communication, that can be conveyed via cinema, television, videos and websites. How does multimodal communication convey emotion and how can we detect and capture them? The visual, adding image to oral communication, in a certain sense enhances the emotional effect, even if it does not automatically add more information in comparison to only audio cues (Esposito, 2007). Furthermore, trailers and films present a narrative structure which influences the emotional impact of these iconic products. Linguistics has provided a very effective strategy to analyse this kind of cultural artefacts. Bortoluzzi's paper shows very clearly the process of social construction of emotion as it is embedded in the documentary movie trailer for 'An Inconvenient Truth: A Global Warning' by Al Gore, former vice president of the USA, who is now engaged in the international mobilisation against global warming. The methodology applied in the study to analyse the role of emotion in Al Gore's trailer is systemic functional grammar in a social-semiotic perspective. The results show that trailers like this, although they appear to the general public to have a neutral objective, instead play on embedding ideological tenets and constructing multiple identities that found their persuasive capacity on emotion; in this sense emotion is the key to the narrative.

At the end of this section the chapter by Tom Denison, Stefanie Kethers and Nicholas McPhee, called 'Implementating E-Research Environments: The Importance of Trust', looks into the emotional issue of e-science. E-science is a new field of research which investigates the influence of the internet in the process of science production. But, as Denison, Kethers and McPhee show, the development of e-science is also affected by the emotional issue that involves all computer-mediated communication: trust. There is a huge literature about the limits and constraints that the lack of trust produces in interlocutors. However, this literature mainly regards trust connected with the constrained negotiation that usually takes place among interlocutors in the internet, when they do not know each other previously. Trust, which is, according to Simmel (1983), one of the most synthetic strengths in society, is in general negotiated in body-to-body encounters and relationships through the exercise of the emotional intelligence. Many data about the interlocutors are processed at the same time: how persons present themselves, how they behave, how they deal with the social context. Without this data, one would find little is known about whom to trust. In fact: '[...] involuntary emotion signals provide some of the most, reliable information about people's characters' (Evans, 2001, p. 43). Denison et al., in discussing emotion in research about the on-line environment, explore the core of the emotional problems associated with e-research. This is a typical case, albeit rarely researched, in which the immediate content of the relationship between researchers leaves aside their reciprocal knowledge. Whilst the importance of trust can be anticipated, Denison et al develop this exploration by analysing the 'non-technical needs' of the researchers. The value of this research is the way in which trust is measured. These results give us a greater understanding of the complexity of communication inside the scientific knowledge production process and explain why the development of cyberinfrastructure for research has limitations for its implementation, limitations which above all are emotional.

## Theme 3: The Emotional Investment Users put into ICTs

The mobile phone has an omnipresence worldwide that would appear to be different from other ICTs and here we learn about the ways that people have appropriated mobile phones into their everyday lives to support, or indeed become part of their identity. But the emotional investment in ICTs is not limited to the mobile phone and the everyday life: it includes also the work environment and the intranet, as Giuseppina Pellegrino underlines in her exploration of ICT users in some business environments.

Jane Vincent's chapter, 'Emotion, My Mobile, My Identity', which opens this third section, observes that the mobile phone capabilities that started from a few hundred mobiles twenty or so years ago have now been adopted by half of humankind. The speed of diffusion highlights the question of emotion; domestication of the mobile phone becomes an issue and its social representation a problem. New generations are now mobile native with the concept of mobile phone communication being integrated into people's lives from birth. Is the mobile phone becoming invisible as has happened to the fixed telephone? Drawing on her growing body of research on the topic conducted over several years, Vincent explores the emotional attachment some people have to their mobile phone. She shows why it is important to work on the emotional attachment to this technology on a macro-sociological level, in order to understand the quality of the relationships passing through this device. As Miller shows in his book *The Comfort of Things* (Miller, 2008, p. 1), the commonplace according to which our relationships with things come at the expense of our relationships to people must be rejected, since: '[...] usually the closer our relationships are with objects, the closer our relationships are with people'.

Giuseppina Pellegrino's chapter, 'Learning from Emotions Towards ICTs: Boundary Crossing and Barriers in Technology Appropriation', takes us smoothly into an important setting, different from the domestic one, that of business. Industrialized countries are moving from a manufacturing economy to a service-oriented economy, in which the added value is represented by the capacity to manage relationships and thus emotion. In fact at the heart of service are relationships: interpersonal, intergroup and interdepartmental relationships (Bliss, 2008). In her chapter Pellegrino analyses the importance of the role of emotion in the implementation of socio-technical systems, in particular the intranet, and specifically in the co-construction of the relationship between managers and technology in working places. She looks at the appropriation and social learning process of the technologies of communication and information used inside the two opposite fronts, in favour (hopes) and against (fear, horror and so on), exploring the conceptual aspects of the boundaries between these various emotions and technologies. As Goleman (1998), showed, emotional intelligence in working settings is a combination of skills, such as a person's ability to manage and monitor their own emotion, to correctly gauge the emotional state of others and to influence opinions. In particular, he describes a model of five dimensions: self-awareness, self-management, motivation, empathy and social skills. Pellegrino's analysis situates itself in the second dimension, that of self-regulation, which, among the other skills, includes the capacity to handle change (adaptability) and to be comfortable with novel ideas, approaches, and technologies (innovation). She investigates specifically the emotional reaction of managers to the introduction of new technologies.

## Concluding Thoughts

The editors of this book propose and discuss the notion of electronic emotion and root it in a theoretical framework with the aim to improve the understanding of the relationship between emotion and ICTs. They also build a three-part model to provide a structure, hopefully useful, in which to situate the research on emotion and ICTs carried out up to now. With this book the editors provide a multi-disciplinary international collection of new research on human emotion and ICTs by leading academics. It is the beginning of an epidemiology of studies firstly on the emotional experience in the practices of use (and non-use) of all the information and communication technologies, secondly on the emotion mediated by the content ICTs convey and thirdly on the emotional investment that users put into the ICTs and the devices, with regard to the self and personal identity. With this volume we aim to provoke a rethink by scholars who work on ICTs, so that they take into account in their approaches and empirical research, that emotion is a fundamental component of humans' attitudes and behaviours. We further hope that this volume could be a turning point for the research on the use of information and communication technologies. At the same time, we seek to open a new field of research in the studies of ICTs that will look at electronic emotion from different theoretical and disciplinary approaches. In this way we aim, not only to systematize the debate and overcome the current fragmentation and dispersion of the scientific reflection on the mediation of emotion via information communications technologies but also, to solicit further research on this topic.

## References

Alquati, R., 'Composizione del capitale e forza lavoro alla Olivetti', *Quaderni Rossi* (3), 1970: 119–85.

Andersen, P. A. and Guerrero, L. K. (eds), *Handbook of communication and emotion: Research, theory, applications, and contexts*, San Diego, CA: Academic Press, 1998.

Barbalet, J. M., *Emotions, Social Theory and Social Structure*, Cambridge: Cambridge University Press, 1998.

Bateson, G., *Steps to an Ecology of Mind: Collected Essays in Anthropology, Psychiatry, Evolution, and Epistemology*, Chicago, IL: University of Chicago Press, 1972.

Benjamin, W., *The Work of Art in the Age of Mechanical Reproduction*, 1936. Available at <http://www.marxists.org/reference/subject/philosophy/works/ge/benjamin.htm>.

Blackburn, S., *Ruling Passions*, Oxford: Clarendon Press, 1998.

Bliss, S. E., *The Affect of Emotional Intelligence on a Modern Organizational Leader's Ability to Make Effective Decisions*, paper retrieved 20 July 2008 at <http://eqi.org/mgtpaper.htm>.

Bondi, L., 'Empathy and Identification: Conceptual Resources for Feminist Fieldwork', *ACME: International Journal of Critical Geographies* (2), 2003: 64–76.

——, 'The Place of Emotions in Research: From Partitioning Emotion and Reason to the Emotional Dynamics of Research Relationships', in Davidson, E., Bondi, L. and Smith, M. (eds), *Emotional Geographies*, London: Ashgate, 2005.

Brants, K., 'Who's Afraid of Infotainment?', *European Journal of Communication* (3), 1998: 315–35.

Carey, J. W., *Media, Myths, and Narratives*, Newbury Park: Sage, 1988.

Castells, M., Fernández-Ardèvol, M., Qiu, J. L. and Sey, A., *Mobile communications and society: A global perspective*, Cambridge, MA: MIT Press, 2007.

Collins, R., *Conflict Sociology. Toward an Explanatory Science*, New York, NY: Academic Press, 1975.

——, 'Stratification, Emotional Energy, and Transient Emotions', in Kemper, T. D. (ed.), *Research Agendas in the Sociology of Emotions*, Albany, NY: State University of New York Press, 1990.

Contarello, A., Fortunati, L., Gomez Fernandez, P., Mante-Meijer, E., Vershinskaya, O. and Volovici, D., 'ICTs and the human body: An empirical study in five countries', in Loos, E., Haddon, L. and Mante-Meijer, E. (eds), *The Social Dynamics of Information and Communication Technology*, Aldershot: Ashgate, 2008.

Denzin, N. K., *On Understanding Emotions*, San Francisco, CA: Jossey-Bass, 1984.

——, 'On Understanding Emotion: The Interpretative-Cultural Agenda', in Kemper, T. D. (ed.), *Research Agendas in the Sociology of Emotions*, Albany, NY: State University of New York Press, 1990.

Durkheim, E., *Les formes élémentaires de la vie religieuse. Le système totémique en Australie*, Paris: F. Alcan, 1912.

Eco, U., *Apocalittici e integrati: comunicazioni di massa e teorie della cultura di massa*, Milan: Bompiani, 1964.

——, *A Theory of Semiotics*, Bloomington: Indiana University Press, 1976.

Elias, N., *Uber den Prozess der Zivilisation*, Frankfurt: Surkamp 1969–80, 1939.

Elkman, P., *Emotion in the Human Face*, Cambridge: Cambridge University Press, 1982.

Esposito, A., 'The Amount of Information on Emotional States Conveyed by the Verbal and Nonverbal Channels: Some Perceptual Data', in Stilianou, Y. et al. (eds), *Progress in Nonlinear Speech Processing. Lecture Notes in Computer Science* (4391), Berlin: Springer Verlag, 2007.

Evans, D., *Emotion. A Very Short Introduction*, Oxford: Oxford University Press, 2001.

Ferrarotti, F., *Macchina e uomo nella società industriale*, Rome: ERI, 1970.

Flam, H., '"The Emotional Man" and the Problem of Collective Action', *International Sociology* (5), 1990: 35–56.

Fortunati, L., *L'arcano della riproduzione*, Venice: Marsilio, 1981 (English tr. *The Arcane of Reproduction*, New York: Autonomedia, 1995).

——, 'I mostri nell'immaginario', Milan: Angeli, 1995a.

—— (ed.), *Gli italiani al telefono*, Milan: Angeli, 1995b.

——, 'User Design and the Democratization of the Mobile Phone', *First Monday* (7): <http://firstmonday.org/issues/special11_9/fortunati/index.html>, 2006.

Fortunati, L., Katz, J. E. and Riccini, R. (eds), *Mediating the Human Body: Technology, Communication and Fashion*, Mahwah, NJ: Erlbaum, 2003.

Fortunati, L., Lee, F. and Lin, A., 'Introduction to the Special Issue on Mobile Societies in Asia-Pacific', *The Information Societies* (3), 2008: 135–9.

Fortunati, L. and Manganelli, A., 'La comunicazione tecnologica: Comportamenti, opinioni ed emozioni degli Europei', in Fortunati, L. (ed.), *Telecomunicando in Europa*, Milan: Angeli, 1998.

Fortunati, L., Vincent, J., Gebhardt, J., Petrovčič, A. and Vershinskaya, O., *Interacting in Broadband Society*, Berlin: Peter Lang (2009).

Freud, S., 'Beyond the pleasure principle' (1920), in Strachey, J. (ed.), *The Standard Edition of the Complete Psychological Works of Sigmund Freud*, Vol. 18, London: Hogarth, 1955.

Fussell, S. R. (ed.), *The Verbal Communication of Emotion: Interdisciplinary Perspectives*, Mahwah, NJ: Erlbaum, 2002.

Fussell, S. R. and Moss, M. M., 'Figurative Language in Emotional Communication', in Fussell, S. R. and Kreuz, R. J. (eds), *Social and Cognitive Approaches to Interpersonal Communication*, Mahwah, NJ: Erlbaum, 1998.

Gallino, L., 'Stratificazione Sociale', entry of *Manuale di Sociologia*, Turin: Utet, 1978.

Gershenfeld, N., *When Things Start to Think*, New York, NY: Henry Holt and Company, 1999.

Goffman, E., 'Embarrassment and Social Organizations', *American Journal of Sociology* (62), 1956: 264–71.

——, *Encounters. Two Studies in the Sociology of Interactions*, Indianapolis, IN: Bobbs-Merril, 1961.

——, *Strategic Interactions*, Philadelphia, PA: Philadelphia University Press, 1969.

Goleman, D. P., *Emotional intelligence*, New York, NY: Bantam Books, 1995.

——, *Working with Emotional Intelligence*, New York, NY: Bantam Books, 1998.

——, *The Emotionally Intelligent Workplace*, San Francisco, CA: Jossey-Bass, 2001.

——, *Social Intelligence: The New Science of Social Relationships*, New York, NY: Bantam Books, 2006.

Griffin, D. R., *The Reenchantment of Science: Postmodern Proposals*, Albany, NY: State University of New York Press, 1988.

Haddon, L., 'Il controllo della comunicazione. Imposizione di limiti all'uso del telefono', in Fortunati, L. (ed.), *Telecomunicando in Europa*, Milan: Angeli, 1998.

Haddon, L., Mante, E., Sapio, B., Kommonen, K-H., Fortunati, L. and Kant, A. (eds), *Everyday Innovators. Researching the Role of Users in Shaping ICTs*, Dordrect: Springer, 2005.

Hammond, M., 'Affective Maximization: A New Macro-Theory in the Sociology of Emotions', in Kemper, T. D. (ed.), *Research Agendas in the Sociology of Emotions 1990*, Albany, NY: State University of New York Press, 1990.

Hardt, M. and Negri, A., *Empire*, Cambridge, MA: Harvard University Press, 2000.

Harré, R. and Gillett, G., *The Discursive Mind*, Thousand Oaks, CA: Sage, 1994.

Heise, D. R., *Understanding Events: Affect and the Construction of Social Action*, New York, NY: Cambridge University Press, 1979.

——, 'Modelling Symbolic Interaction', in Lindenberg, S., Coleman, J. S. and Nowak, S. (eds), *Approaches to Social Theory*, New York, NY: Russel Sage Foundation, 1986.

Hochschild, A. R., 'Emotion Work, Feeling Rules, and Social Structure', *American Journal of Sociology* (85), 1979: 551–75.

——, *The Managed Heart: The Commercialization of Human Feeling*, Berkeley and Los Angeles, CA: University of California Press, 1983.

——, *The Second Shift: Working Parents and the Revolution at Home*, New York, NY: Viking Penguin, 1989.

——, *The Time Bind: When Work Becomes Home and Home Becomes Work*, New York, NY: Metropolitan Books, 1997.

——, 'Love and Gold', in Newman, D. M. and O'Brien, J. A. (eds), *Sociology: exploring the architecture of everyday life*, London: Sage, 2002.

——, 'On the Edge of the Time Bind: Time and Market Culture', *Social Research* (2), 2005: 339–54.

Holland, J., 'Emotions and Research', *International Journal of Research Methodology* (3), 2007: 195–200.

Ito, M., Okabe, D. and Matsuda, M. (eds), *Personal, portable, pedestrian: Mobile phones in Japanese Life*, Cambridge, MA: MIT, 2005.

Jung, C. G. and De Laszlo, V. S., *Psyche and Symbol: A Selection from the Writings of C. G. Jung*, Garden City, NY: Doubleday, 1958.

Lazzarato, M., *Lavoro immateriale* [Immaterial labour], Verona: Ombre corte, 1997.

Ling, R., *New Tech, New Ties: How Mobile Communication Is Reshaping Social Cohesion*, Cambridge, MA: MIT Press, 2008.

Ling, R. and Pedersen, P. E. (eds), *Mobile communications. Re-negotiation of the Social Sphere*, London: Springer, 2005.

Longo, G. O., 'Lo scenario: uomo, tecnologia e conoscenza', in Apuzzo G. M., Araldi S. and Barbieri Masini, E. (eds), *Uomo, tecnologia e territorio*, Trieste: Area Science Park, 2003.

Lutz, C. and White, G. M., 'The Anthropology of Emotions', *Annual Review of Anthropology* (15), 1986: 405–36.

Kay, M. and Svasek, M., *Mixed Emotions: Anthropological Studies of Feeling*, Oxford, UK and New York, NY: Berg, 2005.

Kemper, T. D., *A Social Interactional Theory of Emotions*, New York, NY: Wiley, 1978.

——, 'How Many Emotions Are There? Wedding the social and Autonomic Components', *American Journal of Sociology* (93), 1987: 263–89.

——(ed.), *Research Agendas in the Sociology of Emotions*, Albany, NY: State University of New York Press, 1990.

Maldonado, T., *Memoria e conoscenza. Sulle sorti del sapere nella prospettiva digitale*, Milan: Feltrinelli, 2005.

Mead, G. H., *Mind, Self, and Society*, Chicago, IL: University of Chicago Press, 1934.

——, *The Philosophy of the Act*, Chicago, IL: Chicago University Press, 1938.

Miller, D., *The Comfort of Things*, Cambridge: Polity Press, 2008.

Mosco, V., *The Digital Sublime. Myth, Power, and Cyberspace*, Cambridge, MA: MIT Press, 2004.

Mumford, L., *The Myth of the Machine, Vol. I, Technics and Human Development*, New York, NY: Harcourt, Brace and World, 1967.

——, *The Myth of the Machine, Vol. II, The Pentagon of Power*, New York, NY: Harcourt, Brace Jovanovich, 1970.

Niedenthal, P. M., Krauth-Gruber, S. and Ric, F., *Psychology of Emotion. Interpersonal, Experiential, and Cognitive Approaches*, New York, NY: Psychology Press, 2006.

Nussbaum, M. C., *Upheavals of Thoughts. The Intelligence of Emotions*, Cambridge: Cambridge University Press, 2001.

Nye, D., *Technology Matters. Questions to live with*, Cambridge, MA: MIT Press, 2007.

O'Connell, M. R., *Blaise Pascal: Reasons of the Heart*, Grand Rapids, MI: Eerdmans, 1997.

Parrott, G., *Emotions in Social Psychology*, New York, NY: Psychology Press, 2000.

Piaget, J. and Inhelder, B., *La psychologie de l'enfant*, Paris: P.U.F., 1966.

Rice, R. E. and Love, G., 'Electronic Emotion: Socioemotional Content in a Computer-mediated Network', *Communication Research* (14), 1987: 85–108.

Rime, B., Corsini, S. and Herbette, G., 'Emotion, Verbal Expression, and the Social Sharing of Emotion', in Fussell, S. R. (ed.), *The Verbal Communication of Emotion: Interdisciplinary Perspectives*, Mahwah, NJ: Erlbaum, 2002.

Salovey, P. and Mayer, J. D., 'Emotional intelligence', *Imagination, Cognition, and Personality* (9), 1990: 185–211.

Scott, J. and Marshall, G., 'Stratification', entry of *Dictionary of Sociology*, Oxford: Oxford University Press, 2005.

Short, J., Williams, E. and Christie, B., *The Social Psychology of Telecommunications*, London: Wiley, 1976.

Simmel, G., *Soziologie. Untersuchungen über die Formen der Vergesellschftung*, Berlin: Duncker and Humblot 1983.

Simon, H., 'Motivational and Emotional Controls of Cognition', *Psychological Review* (74), 1967: 29–39.

Smith, A., *The Theory of Moral Sentiments*, 1759. Retrieved 1 August 2008 at: <http://www.marxists.org/reference/archive/smith-adam/works/moral/part07/part7d.htm>

Stets, J. E. and Turner, J. H. (eds), *Handbook of the Sociology of Emotions*, New York, NY: Springer, 2006.

Tarantino, M., 'Techno-myths and techno-faith? Ideas for a cross-cultural analysis', paper presented at the international conference *The Role of New Technologies in Global Societies*, Hong Kong, 2008.

Thoits, P. A., 'The Sociology of Emotions', *Annual Review of Sociology* (15), 1989: 31–42.

——, 'Identity Structures and Psychological Well-Being: Gender and Marital Status Comparisons', *Social Psychology Quarterly* (3), 1992: 236–56.

Vincent, J., 'Are people affected by their attachment to their mobile phone?', in Nyiri, K. (ed.), *Communications in the 21st Century: The Global and the Local in Mobile Communication: Places, Images, People, Connection*, Vienna: Passagen Verlag, 2005.

——, 'Emotional Attachment to Mobile Phones', *Knowledge Technology, and Policy* (19), 2006: 39–44.

Vincent, J. and Harris, L. J., 'Effective Use of Mobile Communications in e-government – How do we reach the tipping point?', *Information, Communication and Society* (11), 2008: 395–413.

Weber, M., *From Max Weber: Essays in Sociology*, translated and edited by Gerth, H. H., Wright Mills, C., New York, NY: Oxford University Press, 1946.

Wright Mills, C., *White Collar: The American Middle Classes*, New York, NY: Oxford University Press, 1951.

# Theme 1

# The Emotion ICT Users Feel when Using, or Not Using, Different Devices

# Old and New Media, Old Emotion

LEOPOLDINA FORTUNATI

## Introduction

In the first part of this chapter I discuss the reasons why it is necessary to look at the emotional sphere in order to understand the complexity of the relationship between human beings and the communication and information technologies (ICTs). In particular, I choose as a point of reference the three dimensional 'model of action' (normative, emotional and rational) which has been proposed recently in the sociological debate to understand the current complexity of social action (Turnaturi, 1995; Flam, 1990). I then examine the results of some quantitative and qualitative research carried out in Italy (Fortunati, 1995, 1998), regarding the emotional impact of the telephone ring, the telephone voice and the answering machine. In the final section I describe the map of emotions provoked by the main ICTs in Europe (fixed and mobile telephone, television, computer, internet, fax, answering machine, radio) focusing on five European countries – France, Italy, Germany, Spain and the United Kingdom. Finally I explore the main differences between these countries with regard to the ownership and consumption of ICTs in relation to emotion.

## Emotion and Communication and Information Technologies

The communicative environment is inhabited by individuals about whom it is necessary to reconstruct the complexity of social action with regard to communication in general and specifically about communicative technologies. Nowadays it is an increasingly shared opinion that the complexity of social action cannot be limited to the classical model of social action at the base of which there is the rational 'man'. Although referred to Max Weber (1922), this classical model is in fact a vulgate and thus is not so faithful to Weber's thinking in which he distinguishes four categories of social action. According to Weber, social action is determined by instrumental rationality, by ethical rationality, *affection* and adherence to tradition or acquired habits. Clearly in Weber's model emotion is one of the key elements to be considered in the analysis. In effect, Weber's model is also reconfirmed by Piaget's theory (1950). In the perspective of the Swiss scholar, emotion and feelings express the interests and values of actions, while intelligence inspires their structure. The emotional as well the intellectual life are a continuous adaptation to the changes of the daily experience in which assimilation and accommodation are interdependent. Emotions and intelligence give birth each to a specific system of schemes, which set up the complementary aspects of a unique reality.

To remedy the limitations of the many analyses of the social action understood only as a rational behaviour, there have recently been several attempts to elaborate a more complex model. The most interesting attempt is that which approaches social action in its three dimensions: normative, emotional, and rational. This model assumes social phenomena as a multi-dimensional process and takes a point of departure for the analysis of social action the emotional dimension of the individual (Turnaturi, 1995). This three-dimensional model should also be applied in studies in the communication field.

Furthermore, in the current elaboration of individuals assumed as e-actors (Fortunati et al., forthcoming), there is thus the effort to attribute to emotion a primary relevance. These kinds of studies often start from an awareness that emotion has an independent logic and the capacity to structure reality, to build a predictable social reality, and to provide models of elaboration of expectations (Flam, 1990). At the same time, these studies start from the awareness that conducting research on emotion might not be easy, because 'emotion' is difficult to define. There are at least four different theories on emotion connected to motivational, cognitive, psychoanalytic and evolutionary models. According to Plutchik's definition (2002) which embodies the fourth model, there are eight primary emotions divided into four pairs: anger and fear, sadness and joy, surprise and anticipation, disgust and acceptance. According to several scholars, the combination of primary emotions leads to secondary or complex emotion (Johnson-Laird et al., 1988). Secondary emotion is manifold and forms a kind of nuance continuum in which the single emotion has difficulty in being clearly defined. Moreover, feelings, that orient the emotional self towards whatever stimulus, are difficult to investigate: they are not controllable as they cannot be induced under command; they might be ambivalent, as antagonist feelings can coexist in the same person, and they might be inconstant, as feelings can run out, change or transform themselves into their contrary meaning. Feelings that are manifold, layered and ambivalent create an empathetic attitude (positive or negative) that shapes the substance of each relationship.

Apart from these general difficulties, the relationship between individuals and machines is also characterized by specific difficulties, given the nature of the latter. Some research highlights a certain emotional underdevelopment towards technologies in comparison to other objects – such as clothing.[1] Other research, however, underlines that individuals are able in time to humanize each object, ICTs included. E-actors anthropomorphize technological artefacts and attribute to them typically human characteristics and qualities (Contarello et al., 2008), with the consequence that they can recreate, in the relationship with objects, those satisfactions or frustrations which mark the interpersonal relationships (Douglas and Isherwood, 1996; Codeluppi, 1995). So, the relation between individuals and machines needs a certain time to develop an emotional intensity.

Emotion is fundamental both in the domestication of a technological artefact in daily life and in its social representation at the level of social thinking as well as at the level of cultural debate. Several works of Vincent are seminal on this issue (2003, 2005, 2006).

---

[1] In a qualitative research carried out on trousers as bins of technologies, in a 10-point scale that aimed to measure the affective attachment to different fashion objects and technologies by interviewees, the mobile received an average assessment of 2.68, which is rather low (Fortunati and Tassile, 2006).

Emotion also occurs in several studies carried out about social representations of telecommunications (Fortunati and Manganelli, 2007) and of the new media (Fortunati and Contarello, 2005; Contarello et al., 2007). These studies show the role of emotions in the cognitive integration of the technologies of communication and information that also reflects the cultural debate that takes place about them. In Western cultures the epistemological debate about technology has been traditionally dominated (at least in the last two centuries) by strong emotional reactions. We have been accustomed by a cultural tradition that sets humanism against technical, machine against human, humanistic civilisation against mechanical civilisation (Ferrarotti, 1970, p. 35). This contraposition is not completely unjustified, but, as Ferrarotti argues, we can not afford to go in the wrong track because of its rhetoric. It is worth recalling that the anti-machinism has always found its most convinced protagonists in persons who do not have any familiarity with machines. The true motivation of this attitude has been well understood by Veblen (1914), who argued that the great guilt of technology was born with the ignominious mark of the manual contamination, the sign of servile labour. For this reason, the anti-machinism was born as a bourgeois myth which shows the inadequacy of the bourgeoisie in respect to the capitalist system which is, however, managed by it. This bourgeois culture is still a lonely and narcissistic culture which excludes labour and its techniques as servile. However, the anti-machinism is not a prerogative only of the bourgeoisie. There are many examples of anti-machinist behaviour as forms of struggle or resistance by workers (luddites, saboteurs, hackers and so on), who fight against a techno-organisational system that expropriates them of the control of their movements and gestures. But it was the leisure class, continued Veblen, that has often theorized the anti-machinism, since it ended by finding itself isolated and superfluous in a technological environment that defined the modern world as rational and secular. This resistance on principle that Veblen observed is also relevant in the present day.

In Europe as well as in China the most proactive users of the new media have been the middle classes, not the upper classes who arrived later to the appropriation of these technologies (Fortunati and Manganelli, 1998; Fortunati et al., 2008). The anti-machinism is connected to a mood, it is an emotional reaction more or less mediated intellectually (Mournier, 1949). Longo (2003) argues that even the analyses about technology conducted in the twentieth century were often emotional, if not visceral, being based on deprecation or enthusiasm. There were in fact intellectuals who saw as the social consequences of the advent of technology only the loss of rationality, the disappearance of the meaning, the premise for a political and ideological totalitarian regime and the tyranny of consumerism. There were other intellectuals, on the contrary, who identified as social consequences of the advent of technology only positive aspects such as the increase of well-being and the development of exchanges and opportunities. These two opposite points of view are both emotional, but fortunately they are increasingly less present in the social debate about technology, that nowadays seems more balanced since it metabolizes both of them. The renewed interest towards the emotional impact of technologies on e-actors is also due to two practical reasons: first, emotions are able to influence the preference for a single device and its use and second, emotions are able to emphasize the strong and weak points of the communication and information technologies themselves.

There are mainly three areas of the relationship between emotions and ICT that have been investigated until now: the emotional aspects of the ICT content, the emotional reactions to ICTs and the emotional investment e-actors develop towards their devices. Of these different areas of the debate, the research I will expand upon here deals only with the second area.

## Aim and Method

I examine here the most salient results of some research projects regarding the social use of ICTs, but only with regard to those aspects connected with the emotional reactions to the communication and information technologies. There are six research projects involved in this analysis on emotion and ICTs. First is a telephone survey carried out in Italy in 1993 (4,130 questionnaires) that investigated mainly the telephone world of the fixed and mobile phone, and the answering machine. Next a body-to-body survey was conducted that involved 101 preschool and 202 primary school children. Third a qualitative research study on youth behaviour related to the fixed telephone world in which 25 students filled in a diary for a week in which they had to record all their incoming and outgoing calls. Fourth is another qualitative research study in which the emotional components of 863 calls made and received by the members of 12 families were investigated and finally a qualitative research study in which 100 university students responded to a semantic differential with 8 stimuli, namely the telephone, yourself on the phone, the ring, the local call, the long distance call, the recipient, the telephone conversation and the mobile. Fifth is again a qualitative research study consisting of 30 interviews with a convenient sample of 30 mothers having children below 3 years old. Finally a telephone survey carried out in 1996 in five European countries – France, Italy, Germany, Spain and the United Kingdom, (6,609 questionnaires) that studied telephone, mobile phone, television, radio, computer, internet, fax and answering machine. In the conclusion I briefly discuss the main problems which remain open on a methodological level in these research projects.

## Results

### *The Telephone Ring*

Katia Kraly[2] underlines that the ring is a sound produced which aims to demand attention, similar to the doorbell. In fact, it is made up of high frequencies which are able to penetrate sound strata. To verify the emotional impact of the ring we asked 4,130 respondents (a representative sample of the Italian population) to answer the question: what is the most frequent emotion you feel when you hear the telephone ring? This question was inspired by previous research carried out by Singer in Canada

---

2    Katia Kraly is a musicologist who was interviewed as an expert on the topic of ring and
      voice.

(1981), in which the ring was not perceived as disturbing by almost half of respondents, was indifferent for 10 per cent of them, annoying 15 per cent and bothering 9 per cent. In Italy, the most frequent emotional reaction to the ring was indifference, then curiosity, annoyance and irritation, anxiety and apprehension, pleasure to speak to somebody, pleasure to be called and relief (Table 1).

| | Gender | | | | A.V. | % |
|---|---|---|---|---|---|---|
| | F | | M | | Total | |
| Emotions | A.V. | % | A.V. | % | A.V. | % |
| Indifference | 443 | 21.05 | 595 | 30.05 | 1,038 | 25.41 |
| Curiosity | 490 | 23.28 | 412 | 20.81 | 902 | 22.08 |
| Annoyance | 314 | 14.92 | 471 | 23.79 | 785 | 19.22 |
| Anxiety | 397 | 18.86 | 221 | 11.16 | 618 | 15.13 |
| Pleasure to speak to somebody | 289 | 13.73 | 170 | 8.59 | 459 | 11.24 |
| Pleasure to be called | 109 | 5.18 | 63 | 3.18 | 172 | 4.21 |
| Relief | 48 | 2.28 | 33 | 1.67 | 81 | 1.98 |
| Other | 15 | 0.71 | 15 | 0.76 | 30 | 0.73 |
| Total* | 2,105 | 100.00 | 1,980 | 100.00 | 4,085 | 100.00 |

Table 1: Emotions engendered by the telephone ring by gender.
* 45 respondents did not answer to this question (A.V.: Number of Respondents).
Source: (Fortunati, 1995).

While in the answers of the Canadian respondents emotional reactions to the ring seemed to be connected mainly to an instrumental perception of telephone calls, those of the Italian respondents were more connected to the relational aspects of telephone calls. Other results suggest that the imagination connected to the ring is also linked to individuals' moods and expectations. In fact a research study carried out by AT&T shows that elderly people do not appreciate the telephone ring because they are afraid that it brings bad news. To other subjects, the ring gives instead a sense of security as it results from several oral stories reported by Fischer (1994, p. 269). In any event the ring, being a sound with a strong mediating function, generally has an effect that is much more binding on the imagination than a musical phrase.

To understand further the emotional reactions to the ring, I administered a semantic differential to 100 university students. In this research the telephone ring was perceived as strong, active, dynamic and speedy. This meant that it represented itself in an appropriate way with regard to its purpose, which was to push the user to answer to the phone. The ring was more strongly felt by men than by women, it being more pleasant and less intrusive for the latter than for the former. Gender differences emerged also in respect to the three dimensions that arose from the factor analysis: the ring recalls more to women than to men 'a pleasant intimacy and opening to the world' and 'an appeal to well-being and comfort'. Both men and women, instead, live with the ring as a dynamic factor. From another qualitative research study consisting this time of 30 interviews with mothers having children below 3 years old, curiosity emerged as the most frequent emotional reaction of their children to the ring. Finally, in the quantitative research study with 101 children of a preschool and 202 children of two primary schools (one urban and the other rural) the main emotional reactions to the ring were in the first place curiosity and then pleasure and amusement. In conclusion, the emotion engendered by the ring among Italian youth and children is positive and linked to the relational aspects of telephone calls.

## *The Emotional Impact of the Voice or the Return of Eco's Myth*

The telephone voice makes the myth of Eco live again. At the same time, it represents the more or less conscious achievement of the ancient dream or delirium of each gnosis: the spiritualisation of the body (Cacciari rep. in Ronsisvalle, 1988, p. 148). With the telephone an extension of the ear and the voice takes place, which is a kind of extra-sensorial perception (McLuhan, 1981). The voice extends itself over distances never imagined before and this produces a displacement to the ear of what was in reality owned by the eye (Adorno, 1951). The transmitter and the recipient, connected through the network, let their minds wander from the context to enter into a kind of bubble, like that of cartoons, where they create an intimate communication. This established circuit connects and then isolates them in an intimate communication, where the voice is not the public one addressed to all, but the private voice. The voice is the terrain of mediation which supports the words and is a primary and indispensable element of the communicative instantaneity. On the telephone we are our voice to the extent that the voice becomes our audio portrait (Friedman and Weiss, 1987). The voice has a specific potency on the telephone, since it is a part that speaks for the whole (Esposito, 2007). A sociology of the voice, however, still remains to be considered (Salazar, 1987, p. 16). This may be because the human voice is situated in the paradox of minimal-ism, since vocal acts are reducible to immaterial exchanges which detract from a determination of value. Instead, the voice has a great function, as it allows us to realize social identification. It is the voice in fact that sets up in a certain sense the order of the world, organising the chaos and regulating social behaviour. On the phone, however, the voice also has a particular effect, because it penetrates directly in the ears and enters the intimacy of the psyche. It is the voice that allows us, amongst other things, to decode the affective impact of telephone calls and offers a very useful material to measure emotion. Bologna (1992, p. 102) argues that the voice timbre can be considered like a: '[...] gesture of the soul'; it is easy to guess, for example, if a person smiles talking on the phone. In an important radio station it has been noticed that when anchors of telephone talk shows smiled when talking, the audience sent in fewer complaints (Scott, 1993, p. 29). The voice is the powerful support that guides us in the meanderings of our and others' emotions. In effect it is an archetypal strength that is supplied by a great creative dynamism. The voice expresses people's mental, symbolic, and affective configuration (Zumthor, 1992, p. 10), although its nature is essentially physical, '[...] since it deals with life and death, breath and sound, and it is emanated by the same organs which preside alimentation and survival' (Bologna, 1992, p. 23).

The voice thus 'speaks' our emotion, offering at the same time the advantage that its emotional modalities can be less controlled than the facial expressions of emotion (Ricci Bitti, 1987, p. 102). Davitz (1964 rep. in Ricci Bitti, 1987, p. 110) has detected that emotion is communicated through the variations of some qualities of the voice, such as the timbre, the tone and the rhythm, independently from the verbal content. So, through the auditory canal it is possible to obtain the dynamic interaction of the vocal characteristics of largeness, highness, timbre and rhythm. Scherer (1983, pp. 147–9) argues that it is even possible to recognize emotion on the base of the voice almost as much as it is on the base of facial expressions. In the case of negative emotions there is even a light advantage in favour of the accuracy of voice recognition. The ranking of emotion is better recognized through the voice: rage, followed by sadness, indifference and, lastly, happiness. Disgust, is instead better recognized through the face (Brighetti et al., 1983, p. 162). There exists, however, strong individual differences in coding and decoding emotion, due to age, gender, personality, as well as to the role played by various communicative modalities. For example, women are more able to decipher the facial expressions of primary emotion, but lose this advantage in the recognition of their vocal expressions (Ricci Bitti, 1987, pp. 113–14).

It is starting from these considerations that I carried out qualitative research on the emotional impact of the telephone voice. I asked 25 students to fill in a diary for one week to write up the emotional characteristics of the voice in their incoming and outgoing calls. The proposed adjectives corresponded to those more frequently indicated in a previous pre-test. These were: confidential, affable, informal, intimate, normal, accomplished, tender, concerned, cold, irritated, cheeky, embarrassed, edgy, amused, aggressive, and upstart. The voice tone in the telephone calls had been defined mainly as confidential. If we take into account that the first argument of telephone conversations is the request or the transmission of information, this means that it is about intimate information. It is also worth noticing that the number of the emotions cited by respondents in the category 'other' is high. They describe various components involved in the communicative process: status of the body connected to emotion (relaxed, sleepy); their intensity (very strong); characteristics of behaviour motivated by them (open, spontaneous, natural, familiar, personal, sincere, persuasive, kind, affectionate, friendly, courteous, fair, interrogative, offish, formal, sharp, sharp-cut, neutral, irritating, distrustful, quiet, calm); aspects of the personality (ready and willing, disposable, ironic, dominating, sharing, lap dog, pleasing), and mental status associated to emotion (surprise). If we want to grade the tones of the telephone voice they are more frequently linked to happiness (very happy, close, glad, euphoric, quizzical, relaxing, interested). Less numerous are the adjectives which refer to sadness (dour, sad, depressed, uncomplaining, dark), even less to fear (clumsy, nervous, annoyed) and only two to rage (angry, enraged). To avoid reading these results naively, however, one should remember that in addition to the problem of social desirability, there is the tendency to remember more and better pleasant things and forget those that are unpleasant. On average, half of our memories are pleasant, 30 per cent unpleasant and the rest neutral (D'Urso, 1988, p. 196). Another point of caution is suggested by the fact that '[...] it seems that the material with highly emotional content can be recalled better after a certain period of time rather than immediately after its presentation' (D'Urso, 1988, pp. 194–5).

Thus, since the students involved in this research had to note down immediately after each call their emotion commentary, this partiality in the memory could also account for the judgements expressed by the students who have filled in the diaries. Finally, the mood of those making the judgements is also important in influencing the assessment of emotions: while people with a normal or slightly elated mood make mistakes of optimism in the assessments of themselves and of the others, sad people seem instead to assess correctly. Through the various caveats in reading these results, one can conclude that from this research it emerges that telephone interlocutors are able to develop a very rich emotional area which is connected to the reciprocal recognition of emotions through the voice. As a consequence, the awareness of the artificiality of this mediated communication weakens following the reiteration of this practice.

## The Answering Machine

The answering machine has been introduced to avoid losing incoming calls when one is absent: it is in fact a device that allows one to maintain one's own communicative reachability. At the same time, deferring to it allows the recipients to overcome their real inferiority, determined by the fact that they must respond to the initiative of an unidentified subject. With the answering machine, the activation of the message is not limited to the person who has dialled the telephone number, but is also extended to the person who receives the call. The answering machine can function as a device for listening incognito and serves also to filter incoming calls (Oldendick and Link, 1994; Haddon, 1998; Katz, 1999). From another point of view, however, the answering machine makes his/her owner 'reachable' around the clock. Also, when one decides to ignore the ring, an unwanted call might remain recorded. It is not possible to pretend not to have received it; the caller knows that the recipient knows. The answering machine refers to a unidirectional communication, in which the temporal unit between the caller and the recipient no longer functions. Between the action of the caller and that of the recipient there is the certainty of coordination, in the sense that the former records a message that he/she knows that the latter will listen to later on. But the construction of the meaning of what is said occurs belatedly and unilaterally, and not between two persons who cooperate to understand each other. The answering machine thus supplies filter and memory functions: the recording of the outgoing and incoming message offers the transaction of a service and not an encounter among communicators. By recording the message, the answering machine transforms the conversation from interpersonal communication into social memory. With its message recoded in the same way for anyone who calls, the answering machine makes communication, or better, the communicative interface rigid and indifferent.

But if we look historically at the answering machine and inside the communication environment, then the function of this device appears very important. The answering machine in fact represented the first mass approach to a tactile voice but, what is more important is that it has spread the habit to interact with a machine. The human-machine interaction would have been much more difficult if people had not passed through the experience of using an answering machine. This is exemplified by the results of some research carried out in Italy that helps to understand the emotional integration of this device at the beginning of its mass diffusion. In the Italian survey of 1993 (Fortunati, 1995), only 6.3 per cent of the 4,130 respondents declared they owned an answering machine. However, the majority of the interviewees stated that at least one time in their life they encountered an answering machine when calling somebody. There were two main emotional reactions to this device: irritation and nothing in particular (Table 2). The remaining answers were distributed among distrust, embarrassment, amusement, curiosity, anxiety and relief. The answering machine frustrates the expectation of one of the recognized advantages of the telephone, which is the immediacy of the retro communication: it is for this reason that the answering machine irritates the person who encounters it. Those who call are showing their disposition to communicate and become disappointed by not finding the wanted interlocutor. The distrust arises from the fact that the caller anticipates a certain trust and intimacy since the content of what will be said will be recorded in a voice-to-voice conversation only in the minds of both interlocutors. A telephone conversation is in fact the triumph not only of the language spoken in freedom, but also of the voice which is volatile, as it does not leave traces behind it. It is this certainty that stimulates attitudes of trust and/or intimacy (Claisse et al., 1985). Finding a machine in the place of a person provokes the opposite reaction: distrust. What will be said will be recorded and so it will remain over time: a lot of people clam up in the face of this possibility.

| Effect | Number of Respondents* | % |
|---|---|---|
| Irritation | 1,397 | 44.10 |
| Nothing in particular | 1,165 | 36.78 |
| Distrust | 296 | 9.34 |
| Embarrassment | 72 | 2.27 |
| I do not know | 69 | 2.18 |
| Amusement | 57 | 1.80 |
| Curiosity | 37 | 1.17 |
| Anxiety | 31 | 0.98 |
| Other | 26 | 0.82 |
| Relief | 15 | 0.47 |
| N.A. | 3 | 0.09 |
| Total* | 3,168 | 100.00 |

Table 2: Emotional effect engendered by the answering machine.
* 962 respondents never found an answering machine.
Source: (Fortunati, 1995).

Embarrassment is also easy to understand: from being disposed to a dialogue one must shift instead to a monologue, which additionally will be recorded. Moreover, speaking to a machine instead of with the expected interlocutor, means in a certain sense the acceptance of a self-degradation that implies a loss of one's own situational self-esteem. This embarrassment occurs in a private instead of a public situation, but it equally has those characteristics that make it a true emotional reaction (Contarello and Zamuner, 1991).

To look more deeply into the emotional reaction towards the answering machine research was conducted with 100 university students involving a semantic differential with ten concepts regarding calls. These were: incoming and outgoing, spoken in Italian and in dialect, local and long distance, addressed to known or unknown recipients, and to which an answering machine or a person answered. From this research it emerged that with an answering machine the call becomes less intimate and liberating. Having to manage the expectation of talking to the person called and instead having to improvise a short monologue, in which one must try to say in few seconds what he/she had the intention to say to the interlocutor, makes the caller more confused, less defined and verified. In addition the caller is of course less calm, since the answering machine still agitates and embarrasses many people. Finally, the call also becomes more exceptional and special, given that at the time this device was still scarcely diffused (6.3 per cent, according to Fortunati, 1995). I studied further the emotional reaction towards the answering machine in another quantitative research, already cited, with 101 children of a preschool and 202 children of two primary schools (the former urban and the latter rural). The majority of the primary schools pupils stated that they felt embarrassed and disappointed. The answering machine tended to scare more girls than boys. Rural children felt more embarrassed than the urban children. The preschool children, instead, tended to appreciate the answering machine and were scarcely embarrassed. Among these children, amusement and curiosity derived from the fact that a 'speaking machine' might appear odd and amusing, especially when one is not used to listening to a recorded voice: '[...] it seemed [to be] my aunt, but it was not my aunt' said a preschool boy. There appears to be different emotional reactions towards the answering machine on the part of preschool children and primary school children. Embarrassment is an unimportant emotion in the world of the first infancy, while for primary school children (both rural and urban) it is the most relevant. The line of demarcation is the awareness of social roles that exist in preschool children in a limited way. These children have few problems of looking good in the face of others and thus they rarely feel embarrassed, while the primary school children often feel embarrassed because they are beginning to be aware of the importance of the social roles. Their embarrassment is also stressed by the fact that they have a crisis about whether they have the level of conversational ability required to cope with the answering machine. Finding an answering machine poses in fact a series of problems to resolve: do I leave a message or ring off? If I decide to speak, what do I say? Thus the main reaction of children aged between 7 and 11 years is embarrassment due to the displacement of an interpersonal relationship by a communication mediated via an artificial dimension – the answering machine.

## Emotion and ICTs in Europe

If in Italy, as one learned above, the studies on the emotional reactions towards ICTs were mainly focussed on the telephony area (telephone and answering machine), at a European level the studies conducted have regard to the entire main ICTs: telephone, mobile, television, radio, computer, internet, fax and answering machine. A telephone survey was administered in 1996 to a representative sample of French, Italian, German, Spanish and British populations: 6,609 respondents. In this research we worked with a list of 17 emotions. This list was obtained starting from the emotions identified in previous research (Singer, 1981; Moyal, 1992; Fielding, 1992; Fortunati, 1995), and then by adding other emotions coming from the content analysis of the answers to 'other' obtained in the pre-test of this survey (administered to 100 respondents). The question was: What is the predominant emotion provoked by the single devices?

## The Map of Emotion and ICTs in Europe

More than half of the answers referred to positive emotions: interest, joy, relaxation, company[3] and so on (see Table 3). In particular, the category of enthusiasm, that on the whole collected few marks, was associated especially with the computer, as well as with the category of interest, that had collected a high number of citations. Curiosity and anxiety, along with satisfaction and company, were connected first of all to the telephone. Irritation, anger and unpleasantness were attributed to the mobile phone. Boredom was associated more frequently to the television. Joy, relaxation and amusement were often linked to the radio. Frustration and embarrassment were attributed to the answering machine, while the indifference or the emotional open-endedness to the fax. The key word that opened the world of emotions engendered by ICTs was 'interest'; that is the capacity that the means of communication has to attract people's attention. This capacity to be charming, however, is not associated with the capacity to maintain attention: think for example of the absent-minded and multi-layered consumption of mass media. This double movement – attracting people's attention and losing it – is what characterizes the main tenet of communication and information technologies. Telephone, radio and television shared an emotion equally relevant: joy. They are the only instruments of communication capable of conveying this emotion and for this reason they are particularly precious within the communicative environment. It is the voice that has this enormous trans-figurative power of annulling the existence of the communicative transformer and to supply the pleasure of the encounter with the other. This, even if it is a little distorted by the artificiality of the instrument and in the case of TV weighed down by the image. The other key word, which concerned again the telephone, television and radio, was company. These devices are considered capable of simulating the human presence. They give the impression of entertaining audiences and this capacity becomes very important in a society that always pushes the process of individualisation and isolation of the individual ahead. ICTs

---

3     Company, as identified in the pre-test, refers the mutual comfort of co-presence.

act as a surrogate of communication and socialisation in co-presence that actually tend to be reduced and imploded. This vicarious function of ICTs fills the artificial spaces that until yesterday were occupied by interpersonal communication or by the absence of communication. But it has, however, a positive sign since it is better to share mediated communication with somebody rather than being completely separated from others. Another emotion, very much connected to the television and the radio, is relaxation. This highlights how much the consumption of these two technologies is focussed at an emotional level: the radio is used as background noise in order to sedate tensions, the television is used in the night to send you to sleep like a sleeping pill. The mobile phone, answering machine, fax and computer were all associated with indifference, disinterest, and nonchalance. In 1996, these were technologies that had still to acquire the power of attraction at mass level and they were not yet part of the daily experience of many people. This explains also the fact that many respondents did not articulate an assessment of the mobile phone, fax and answering machine and declared they felt nothing in particular. On the whole, positive emotions were significantly associated with a high frequency use of communicative technologies.

*Emotion, ICTs and European countries*

Helen Flam (1990) argues, each culture has its emotional profile and sensibility. To understand possible national specificities in the emotional reactions towards ICTs, we built a matrix of frequencies with 18 lines (17 emotions and the answer 'I do not feel anything in particular') and 35 columns (7 devices in the five countries) which had been subjected to factor analysis of correspondences (Benzécri, 1976), and is summarized in Table 3. Of the first two dimensions that emerged (that explain 61.5 per cent of the total inertia), the former highlights an opposition between positive emotions such as company, amusement, joy, relaxation and indifference; the latter sees the opposition between interest and emotional open-endedness (Fortunati and Manganelli, 1998).

On the plan defined by the axes of these dimensions, are situated emotion and the communication and information technologies, according to the origin country of respondents. Telephone and classical mass media in almost all the countries were connected to positive emotions. In particular, radio and television for Britons and Germans and only radio for French, are situated around the emotion concepts of joy and relaxation, but they are on the opposite side in respect to interest and curiosity. This indicates the deep pleasure that Britons and Germans had towards the classical media (radio and television), but also that this pleasure was in some way connoted by a sort of passivity. For Spaniards, radio and TV were essentially amusement and for Italians the radio was also company and the small screen was still able to interest and intrigue. Spaniards and Italians, in comparison to the others, had therefore a different, emotional appreciation of the classical media, which was more relational and informative (company and amusement, interest and curiosity). The telephone was exclusively apparent on the emotional plan by the Italians, for whom it was located in the area of interest and curiosity. All the other telephone devices (fax, answering machines and mobile phone), in relation to the first dimension, were situated at the polarity of indifference. By continuing to examine again the plan defined by the axes of these dimensions, a true split emerges, originated by the second dimension, which sees, on one side Germans and Britons, and in part French and Spaniards, and the other side Italians. The Germans and Britons are associated with open-endedness or absence of emotions towards computer, fax, mobile phone and answering machine, the Italians to interest and curiosity not only for the telephone, but also for the other devices. In particular, the technologies that engendered the interest of Italians more are the computer and the mobile phone.

| Emotions | Television | Telephone | Radio | Mobile phone | Answer machine | Fax | Computer | Total Responses |
|---|---|---|---|---|---|---|---|---|
| Interest | 713 | 948 | 595 | 579 | 708 | 813 | 1,089 | 5,445 |
| Joy | 818 | 843 | 1,407 | 205 | 177 | 125 | 396 | 3,971 |
| Indifference | 343 | 199 | 223 | 712 | 805 | 910 | 559 | 3,751 |
| Relaxation | 1,182 | 201 | 1,201 | 82 | 106 | 54 | 135 | 2,961 |
| Company | 468 | 818 | 729 | 191 | 84 | 52 | 58 | 2,400 |
| Amusement | 480 | 200 | 604 | 99 | 70 | 68 | 382 | 1,903 |
| Satisfaction | 161 | 359 | 199 | 201 | 270 | 268 | 248 | 1,706 |
| Curiosity | 271 | 286 | 142 | 126 | 226 | 171 | 239 | 1,461 |
| Irritation | 165 | 192 | 36 | 327 | 301 | 66 | 79 | 1,166 |
| Enthusiasm | 151 | 192 | 185 | 100 | 73 | 103 | 216 | 1,020 |
| Boredom | 391 | 82 | 69 | 130 | 126 | 79 | 103 | 980 |
| Anger | 117 | 107 | 22 | 254 | 204 | 55 | 74 | 833 |
| Frustration | 79 | 79 | 13 | 99 | 175 | 32 | 72 | 549 |
| Anxiety/stress | 19 | 108 | 6 | 93 | 65 | 32 | 85 | 408 |
| Unpleasantness | 65 | 30 | 8 | 104 | 90 | 30 | 26 | 353 |
| Surprise | 19 | 99 | 20 | 36 | 69 | 44 | 23 | 310 |
| Embarrassment | 8 | 23 | 2 | 26 | 60 | 14 | 22 | 155 |
| Nothing | 634 | 704 | 617 | 2,260 | 2,125 | 2,712 | 1,906 | 10,958 |
| Total | 6,084 | 5,470 | 6,078 | 5,624 | 5,734 | 5,628 | 5,712 | 40,330 |

Table 3: Matrix of frequencies to show emotions and ICTs in five European countries. Summary of responses from France, Italy, Germany, Spain, United Kingdom. Source: (Fortunati and Manganelli, 1998).

At the European level (with the exception of Italy) answering machine and fax were related to the status of accessories, which are the additional parts of the telephone devoted to improve its functioning. No more than accessories, these technologies were considered secondary by Europeans. The suspension of emotional judgement or the indifference and the disinterest towards them should probably be attributed to their scarce diffusion in the domestic sphere. This, however, did not discourage Italians from expressing some interest and curiosity also towards them. To summarize these results, one could say that the emotional relationship with communication and information technologies is quite influenced by culture.

A second analysis of this data examined the emotional framework that ICTs provoke in men and women. From the corresponding analysis two dimensions emerged (explaining 83.2 per cent of the total inertia). As in the previous analysis, the first dimension highlights the opposition between positive emotions – such as joy, amusement, relaxation, company – and indifference or any emotion in particular. The second dimension instead opposes company, joy and relaxation to boredom. The radio for both men and women was associated with positive emotions such as joy, relaxation and amusement. In the polarity of indifference there is only the fax for women. The telephone for men and women (but more for women) is near to company. In the opposite polarity there is, instead, the television that, according to men, is located between amusement and boredom, but more displaced towards this latter. We can conclude that, at the European level, apart from a few differences, the emotional attitude of men and women is not different. Summarising these findings, it emerges that the emotional function that both men and women attributed to the radio was rather similar and positive. Equally positive was the function that both attributed to the telephone, even if for women the telephone is closer to company and seems to allow them to overcome social isolation. Furthermore, in respect of the television, women experienced its small screen more playfully and joyfully than men, who were more ambivalent towards it.

*Emotion According to ICTs Ownership and Consumption of Technologies*

Finally, we studied the emotional impact of communication and information technologies in relation to their ownership and consumption. In short, from these analyses it emerges that having an answering machine at home, a fax, a computer or a mobile phone significantly influences the type of emotion they engender. Those who had a certain familiarity towards these various devices expressed positive emotion more frequently, while those who did not own them, declared indifference or to not feel anything in particular and sometimes boredom. Only in a few cases did domestic familiarity with a communication and information technology originate sensations of indifference or cause absence of emotions. Instead, the emotional open-endedness or indifference concerned the lack of direct knowledge of these devices in the domestic settings. There is only one exception in this framework: the computer, which induced more interest, not in those already familiar with it, but in those who did not have it. This result already prefigured the future diffusion of personal computers in the domestic setting. Summarising these results, it emerges that the medium-high consumption of a communication and information technology usually engendered positive emotion, while it is the low or occasional consumption which provoked indifference or nothing. These results seem to contradict the emotional model of consumption which shows that satisfaction and dissatisfaction depend on the break between one or other of these acts and from the time passing from satisfaction to dissatisfaction (Elster, 1989). It seems that the pleasure people feel when they consume a certain product blunts in time. The data collected in this research shows instead an opposite trend. Boredom, indifference or emotional open-endedness are generally associated with an occasional or low consumption of ICTs. We were, and still are, very far from the addiction level regarding ICTs and from the peak of the curve in regard to the communicative technologies. The wear and tear of time is probably only one of the factors that influence consumption and in general the emotional relationship with ICTs. These factors are still not clear; however, their complexity emerges clearly from the results of the Italian survey already cited. Among those who liked much or very much to make calls de facto only 41.5 per cent made one or two calls a day and among those who liked few or did not at all like to make calls de facto 37.7 per cent declared to make the same number of calls daily. In addition with regard to call durations, those who liked much or very much the telephone were in the majority of cases making short calls. Actually, it is not always possible to choose what one likes to do and to avoid what one does not like to do and, at the same time, it can not be taken for granted that the pleasure that one feels in consuming a certain type of good weakens in time.

In conclusion, in 1996 media represented a positive affective presence in the European daily life, especially the telephone, television and radio. Any criticisms that Europeans have of ICTs did not make a dent in their emotional impact. In particular, Britons and Germans expressed a more passive but deep pleasure towards TV and radio, while Spanish and Italians expressed a more relational and informative appreciation (they also made more live calls to radio and TV). Italians felt interest and curiosity towards the telephone (we recall also that Italians were those who appreciated it most at the European level). In regard to the new media (mobile and computer/internet) and the accessories of the telephone, Italians were the only respondents who declared interest and curiosity also towards them, while all the others perceived them with indifference or open-endedness. These results show a slow emotional appropriation of these technologies by Europeans, with the exception only of the Italians who seemed more rapid in this process. If the domestic familiarity towards communicative technologies and their intense use originates positive emotion in the majority of cases, the increase of their diffusion and of their use should indicate both an increase of positive emotion towards ICTs by Europeans in the future and perhaps, at least in the near term, the beginning of addiction.

The results I have described thus far offer several cues to thinking about emotion and ICTs. However, they should be considered with a certain caution, since a theoretical issue is still open. If we assume Helen Flam's analysis on the different integration of emotion at a cultural level, can this premise jeopardize the validity of a cross-comparability of emotion in different countries? In other words, is emotion really comparable in different countries? Other reasons to be cautious about these results derive from the structural limits of the methods applied: Firstly interviews and questionnaires supply findings not on emotion but on the mental representation of emotion that respondents feel and secondly applying the semantic differential offers ways to understand only the emotional temperature towards the different devices.

## Conclusion

These research projects on the emotional impact of ICT uses have offered food to reflect on the importance of emotion for the humankind in constructing its reality. It was particularly challenging to test the emotional side of the tri-dimensional model of social action –normative, emotional and rational- on the use of ICTs as in their social representation technological artefacts generally result from connection with the coldness of metals and plastics and with the apparent emotional neutrality of normative rationality and scientific knowledge. However, humankind does not hesitate to anthropomorphize the information and communication technologies, but what is more important is that humankind conceptualizes, learns and appropriates technologies through emotions.

As discussed earlier, humankind in the West is finding a way to metabolize ICTs in a more equitable way than in the past, overcoming both fear and over-valuation of these technological artefacts. This review of the research I carried out in the last 13 years highlights a positive but balanced world of emotion around the use of ICTs. Furthermore, the results of these research projects confirm that it is emotion which influences the preference of e-actors for a single device and its use. And it is again emotion which helps us to understand the weak and strong points of the information and communication technologies. If one aspires to deeply understand the practices of use of ICTs and their meaning, then the study of the role of emotion in the social and personal use of ICT emerges as crucial.

To conclude also about the concreteness of the findings I explored in this article, let me summarize them briefly. With regard to the emotional impact of the telephone ring, the answering machine and the telephone voice it emerged that: firstly, Italians have an emotional relationship with the telephone ring that is quite connected to the relational aspects of telephone calls; secondly, the most frequent tone of the telephone voice is confidential and the majority of the others are related to happiness and thirdly the emotional reactions to the answering machine, which represents an important outpost in the realm of human-machine interaction, are instead mainly negative (apart from preschool children who often were amused and curious in front of this device) and attest to the difficult emotional integration of this device at the beginning of its mass diffusion.

With regard to the map of emotion engendered by the telephone, mobile phone, television, radio, computer, internet, fax and answering machine at European level, it emerged that this map was made up of mainly positive emotions, such as interest, joy, relaxation, company and so on. It is also interesting that only positive emotion was significantly associated with a high frequency of ICTs use. Looking at the gender and the different countries, it emerges that the emotional relationship of men and women with communication and information technologies is not fundamentally different, whilst it is quite influenced by culture. Finally, ownership and consumption of ICTs also influence their emotional impact: those who had a certain familiarity towards them (owned and used them consistently) expressed positive emotion more frequently, while those who did not own them or had a low use, declared indifference or nothing in particular and sometimes boredom. This means that the increase of their diffusion and use should indicate a corresponding increase of positive emotion towards ICTs by Europeans in the future and not, at least in the short term, the advent of addiction. At the same time, this result demonstrates that the interpretation given to the anti-machinism inside the theoretical debate I reported at the beginning of the chapter is based on a correct assumption: if you know the machine, you like it.

# References

Adorno, T. W., *Minima moralia. Reflections from a Damaged Life*, London: NLB, 1974 (original edition 1952).

Benzecri, J. P. et al., *L'analyse des données*, Paris: Dunod, 1976.

Bologna, C., *Flatus vocis. Metafisica e antropologia della voce*, Bologna: Il Mulino, 1992.

Brighetti, G., Lavadas, E. and Ricci Bitti, P. E., 'Riconoscimento di emozioni attraverso indici provenienti da un solo canale e indici provenienti da due canali', in Attili, G. and Ricci Bitti, P. E. (eds), *Comunicare senza parole*, Rome: Bulzoni, 1983.

Claisse, G., Vergnaud, T. and Rowe, F., *Téléphone, Communication et Société. Recherche sur l'utilisation domestique du telephone*, Lyon: Cnrs, Entpe, 1985.

Codeluppi, V., *I consumatori*, Milan: Angeli, 1995.

Contarello, A. and Fortunati, L., 'ICTs and The Human Body. A Social Representation Approach', in Law, P., Fortunati, L. and Yang, S., (eds), *New Technologies in Global Societies*, New Jersey, NJ: World Scientific Publisher, 2006.

Contarello, A., Fortunati, L., Gomez Fernandez, P., Mante-Meijer, E., Vershinskaya, O. and Volovici, D., 'Social Representations of ICTs and the Human Body. A comparative study in five countries', in Loos, E., Mante-Meijer, E. and Haddon, L. (eds), *The Social Dynamics of Information and Communication Technology*, Aldershot: Ashgate, 2008.

Contarello, A., Fortunati, L. and Sarrica, M., 'Social Thinking and the Mobile Phone: a Study of Social Change with the Diffusion of Mobile Phones, Using a Social Representations Framework', *Continuum: Journal of Media and Culture*, 21 (2), 2007: 149–63.

Contarello, A. and Zamuner, V., 'Imbarazzo e cultura', in D'Urso. V. (ed.), *Imbarazzo, vergogna e altri afán*, Milan: Cortina, 1991.

Douglas, M. and Isherwood, B. C., *The world of goods: towards an anthropology of consumption: with a new introduction*, London: Routledge, 1996.

D'Urso, V., 'Questioni di metodo e Stati emotivi e memoria', in D'Urso, V. and Trentin, R. (eds), *Psicologia delle emozioni*, Bologna: Il Mulino, 1988.

Elster, J., *Nuts and Bolts for the Social Sciences*, Cambridge: Cambridge University Press, 1989.

Esposito, A., 'The Amount of Information on Emotional States Conveyed by the Verbal and Nonverbal Channels: Some Perceptual Data', in Stilianou, Y. et al. (eds), *Progress in Nonlinear Speech Processing Lecture Notes in Computer Science* (4391), Frankfurt: Springer Verlag, 2007.

Ferrarotti, F., *Macchina e uomo nella società industriale*, Rome: ERI, 1970.

Fielding, G., 'Observations Relating to British Research, Initiative and Needs', in Moyal, A. (ed.), *Research on Domestic Telephone Use*, Melbourne: Circit, 1992.

Fischer, C., *America calling: A social history of the telephone to 1940*, Berkeley, CA: University of California Press, 1992.

Flam, H., 'The Emotional Man' and the Problem of Collective Action', *International Sociology* (5), 1990: 35–56.

Fortunati, L. (ed.), *Gli italiani al telefono*, Milan: Angeli, 1995.

—— (ed.), *Telecomunicando in Europa*, Milan: Angeli, 1998.

Fortunati, L. and Contarello, A., 'Internet-Mobile convergence: Via similarity or complementarity?', *Trends in Communication* (9), 2002: 81–98.

——, 'Social Representation of the Mobile: An Italian Study', in Dong Shing, K. (ed.), *When Mobiles Came: The Cultural and Social Impact of Mobile Communication*, Seoul: Communication Books, 2005.

Fortunati, L. and Manganelli, A., 'La comunicazione tecnologica: Comportamenti, opinioni ed emozioni degli Europei', in Fortunati, L. (ed.), *Telecomunicando in Europa*, Milan: Angeli, 1998.

——, 'The Social Representations of Telecommunications', *Personal and Ubiquitous Computers*, published online 16 January 2007, <DOI: 10.1007/S00779-006-0139-7>.

Fortunati, L., Manganelli, A., Law, P. and Yang, S., 'Beijing Calling ... Mobile Communication in Contemporary China', paper published online by the journal *Knowledge, Technology, and Policy* <http://

www.springerlink.com/content/105285/?Content+Status=Accept ed>, 2008.

Fortunati, L. and Tassile, M., 'I pantaloni come contenitori di tecnologie'. Paper presented at the international conference *Geographies of Wearing*, 4–5 May, Milan, 2006.

Fortunati, L., Vincent, J., Gebhardt, J., Petrovčič, A. and Vershinskaya, O., *Interacting in Broadband Society*, Berlin: Peter Lang (2009).

Friedman, M. and Weiss, J., *Saper usare il telefono*, Milan: Armenia, 1987.

Haddon, L., 'Il controllo della comunicazione. Imposizione di limiti all'uso del telefono', in Fortunati, L. (ed.), *Telecomunicando in Europa*, Milan: Angeli, 1998.

Johnson-Laird, P. N. and Oatley, K., 'Il significato delle emozioni: una teoria cognitiva e un'analisi semantica', in D'Urso, V. and Trentin, R., (eds), *Psicologia delle emozioni*, Bologna: Il Mulino, 1988.

Katz, J., *Connections: Social and cultural studies of the telephone in American Life*, New Brunswick, NJ: Transaction, 1999.

Longo, G. O., 'Lo scenario: uomo, tecnologia e conoscenza', in Apuzzo, G. M., Araldi, S. and Barbieri, Masini, E. (eds), *Uomo, tecnologia e territorio*, Trieste: Area Science Park, 2003.

McLuhan, M., *Understanding Media: The Extensions of Man*, New York, NY: McGraw-Hill, 1964.

Mournier, E., 'La machine en accusation', in Gurvitch, G. (ed.), *Industrialisation et Technocratie*, Paris: Colin, 1949.

Moyal, A. (ed.) *Research on Domestic Telephone Use*, Melbourne: Circit, 1992.

Oldendick, R. W. and Link, M. W., 'The Answering Machine Generation: Who Are They and What Problem do They Pose for Survey Research?', *The Public Opinion Quarterly* 58 (2), 1994: 264–73.

Piaget, J., *Introduction à l'épistémologie génétique*, Paris: P.U.F., 1950.

Plutchik, R., *Emotions and Life: Perspectives for Psychology, Biology, and Evolution* Washington, DC: American Psychological Association, 2002.

Ricci Bitti, P. E., *Comunicazione e gestualità*, Milan: Angeli, 1987.

Ronsisvalle, V., *Hallo 2000. Il telefono nel terzo millennio. Incontri*, Rome: Sarin, 1988.

Salazar, P. J., 'Formes de la voix', *Sociétés* (16), 1987.

Scherer, K. R., 'La comunicazione non verbale delle emozioni', in Attili, G. and Ricci Bitti, P. E. (eds), *Comunicare senza parole*, Rome: Bulzoni, 1983.

Singer, B. D., *Social Functions of the Telephone*, Palo Alto, CA: R. & E. Research Associates, 1981.

Turnaturi, G. (ed.), *La sociologia delle emozioni*, Milan: Anabasi, 1995.

Veblen, T., *The Instinct of Workmanship and the State of the Industrial Arts*, New York, NY: Macmillan, 1914.

Vincent, J., 'Emotion and mobile phones', in Nyiri, K. (ed.), *Mobile Learning: Essays on Philosophy, Psychology and Education*, Vienna: Passagen Verlag, 2003.

——, 'Emotional Attachment to Mobile Phones: An extraordinary relationship', in Hamill, L. and Lasen, A. (eds), *Mobile World Past Present and Future*, London: Springer, 2005.

——, 'Emotional Attachment and Mobile Phones', *Knowledge, Technology, and Policy* 19 (1), 2006: 39–44.

Weber, M., *Economics and Society*, Tübingen: Mohr, 1922.

Zumthor, P., 'Prefazione', in Bologna, C., *Flatus vocis. Metafisica e antropologia della voce*, Bologna: Il Mulino, 1992.

# Mobile Phone Calls and Emotional Stress

JOACHIM R. HÖFLICH

## Introduction

The mobile phone has been used more and more in public spaces in European countries for over two decades and it is often associated with causing emotional disturbance to others in this public sphere (Burkart, 2000). Now taken up by the majority of the populations of many countries the mobile phone is used almost anywhere thus creating more opportunities for these disturbing moments to occur (Fortunati, 1998). It would appear, however, that social practices relating to mobile phone communications are now emerging that accommodate these emotional disturbances. In this chapter I explore these changing practices and the extent of the obtrusiveness of mobile phones through examination of a study I carried out in Germany in which I used the method of ethnomethodological breaching experiments. The idea for this 'breaching study' came about as a result of examining the findings from earlier studies conducted in four European countries (Höflich, 2005c), as well as being informed by the research of others such as Fortunati (2005) and Ling (2002). In this chapter I will discuss these earlier studies as well as the methodological approach and the findings from the breaching experiments. I will show that the 'ringing' of a mobile phone (or whatever sound it makes) may evoke emotional stress not least because of the negative reactions of the people around the recipient, but also because of the emotional reactions by the recipient who anticipates the negative reactions by bystanders.

# The Mobile Phone as an Obtrusive Medium

The mobile phone is part of our every day life. Sometime we think we cannot live without it and sometimes we hate it because of its intrusiveness and obtrusiveness. When others use their mobile phones it can be terribly disturbing, almost as if another person has intruded. Indeed, studies have demonstrated that the mobile phone is seen as a dominant emotional disturbing factor of the public sphere (see, for instance, Fortunati and Manganelli, 1998; Höflich, 2005a). But people also seem to have a certain feeling regarding a sense of place and where is it appropriate to use a mobile phone. However, at certain times people make its use intolerable and at others mostly acceptable (Höflich, 2005b). The degree of disturbance it causes depends not just where it is used but also on the situational circumstances pertaining at the time. For instance, when in a cinema, a theatre, a museum or a church the mobile phone is considered to be much more disturbing than in the streets, in public parks, or in a pedestrian area. Results of an exploratory international study that was conducted from the end of 2002 to the beginning of 2003 illustrated this disturbing use of mobile phones (Höflich, 2005c, p. 129). This study was carried out in Spain, Italy, Finland and Germany and Table 1 shows situations researched in the study where the use of the mobile phone is perceived to be a particular nuisance by the total of 400 respondents we asked in the four countries.

The results of this study showed, in common with several other cross-cultural research projects (see for instance Fortunati, 1998), that there were cultural differences evident in terms of people's assessment of the mobile phone as a disturbing device in different settings. Looking at Table 1 the four types of situations listed with the highest degree of emotional nuisance were similar for all the countries studied. However, for the other locations and situations this was not the case, most particularly in the cases of Spain and Italy where the mobile phone caused much less of a nuisance. The study also showed that despite the fact that mobile phone use in public spaces – in Italy in particular – had become a widespread phenomenon, it was notably the Italian respondents who agreed with the statement 'I feel uncomfortable making a call on my mobile phone if strangers are around me'.

| Places where mobile phone used | Extent of nuisance of the mobile phone (percentage) |
| --- | --- |
| In cinemas, theatres or museums | 92.0 |
| At official events | 91.5 |
| In churches | 89.6 |
| In waiting rooms | 70.8 |
| In restaurants | 57.5 |
| At social events | 47.5 |
| At work | 41.8 |
| Public transportation | 37.5 |
| In pubs or bars | 34.4 |
| At sport events | 29.5 |
| In other people's houses | 27.1 |
| In shops | 25.0 |
| At your home | 18.3 |
| In waiting areas | 14.0 |
| In the streets | 8.1 |
| In public parks | 7.0 |
| In pedestrian areas | 6.0 |

Table 1: Degree of nuisance of the mobile phone according to different places (n=400).

I would assert that this emotional conflict between using one's mobile phone and disturbing others is apparent and suggests that the respective settings are embedded within the overall framework of a 'situational balance' – the handling of which differs from culture to culture, and which may be only understood against that particular background. On the one hand, such emotional disturbances could be seen as a manifestation of an early phase of the diffusion of any new medium that always leads to certain eruptions of communicative practices. On the other hand, disturbances can also be seen as immanent in the use of the mobile phone for you cannot use a mobile phone in a public place without causing some disruption, however small. To a certain extent both points of view are true;

we can expect early phases of new media to always show some eruptions of communication practices or even a certain anomic phase that might cause considerable disruption and upset of existing social practices. But after this instability there follows a process of normalisation, a period of integrating the medium into the practices of everyday life. For instance peoples' use of mobile phones becomes more routine, and as this happens they become tolerated in certain places where previously they may not have been.

As the mobile phone becomes more and more a part of everyday life we can see that an increasing number of people take it for granted as well as the sounds it produces. In the words of Rich Ling:

> Time has shown that the telegraph is forgotten and the two latter innovations have become taken for granted. The ringing of a home telephone is handled with well-oiled routines – if not excitement in the case of lovelorn teens. Switching on a light is only occasionally the focus of a social ritual, as for example when lighting the family Christmas tree or dimming the lights for a 'special' dinner. The mobile telephone may, in all likelihood go on in the same direction (Ling, 2002, p. 83; see also, Ling, 2004, p. 143).

There seems to be, regardless of the media in question, a common process that occurs with its introduction – the three phases of eruption, appropriation, and finally normalisation (although this should not preclude the fact that appropriation in a wider sense is continuously going on).

'Manners matter' one could say and as each new medium begins these phased processes are accompanied by the development of new media etiquette pertinent to that particular medium. For instance in the early years of the telephone a 'telephone courtesy', an 'etiquette of the telephone' was demanded and 'bad telephone manners' were denounced. In 1918, for instance, Mary Mullett wrote in the *American Magazine*:

> The way in which you call a person, or answer him, is a pretty good index of what you are ... The telephone strips the veneer of your manners – if they are veneered (Mullett, 1918, p. 44).

Similar to this example for the land-line telephone a media etiquette was also developed as a consequence of user demands for the telefax, the email and, last but not least, in the case of the mobile phone. Leopoldina Fortunati

(Fortunati, 2005, p. 233) writes that in its early years the mobile phone has been regarded as a status symbol, but one of a vulgar kind. By using – or displaying– the mobile phone people not only demonstrated social status but they also showed bad manners because they did not care about what was going on around them. Step by step the mobile phone has slowly been changing this image, although not completely. In this process of normalisation the old rules of communication that developed from land-line telephone use are changing as new rules emerge that will no doubt finally be taken for granted. However, there will always be some struggle about the rules and this is not least because rules have to be seen in the overall context of communication rather than in isolation. Take for instance the plain old land-line telephone where

> [...] it appears that the rules governing telephone calls cannot be understood unless they are placed within a larger system of interaction which distributes different roles to different means of communication with the other member of the community, a system which one expects to be itself determined by technical and geographical constraints on the one hand, and cultural values and attitudes on the other (Godard, 1977, p. 219).

In the process of appropriation the mobile phone has lost part of its distinguishing role as a status marker but also its discriminating effect as a medium of bad manners. Instead of this other moments became part of a public discourse and the mobile phone is but one medium that intrudes into this complex web of interactions, making it necessary for them to be rearranged (see Ling, 2004, p. 130). Indeed, social arrangements regarding the usage of mobile phones have increasingly emerged, but this does not exclude that even in contexts where the mobile phone usually is accepted a temporal feeling of being disturbed may be evoked. However, if there are disturbing moments people (should have to) try to find solutions and (should try to) make arrangements. This means that they finally have to re-normalize temporally these emotional eruptions to enable their social life to go on.

Nowadays, the mobile phone is an integral part of everyday life. But there are still places and circumstances (such as in the theatre or the church) where the mobile phone will always be an unacceptable disturbance. For instance in places with strong expectations of normative behaviours and the affordance of silence there may be moments when mobile phone use has to be managed. Take the example of a library. A study we conducted at the library of the University of Erfurt (Höflich et al., 2007), indicates that it is not enough to ask people to be considerate of others with regard to the use of their mobile phone, and not to use it in a way that might cause disturbance. Instead of this one has to look more deeply into the situational circumstances. Indeed, it could be shown that generally most people when asked confirmed that the mobile phone is a disturbance or intrusion. But looking more closely we found a difference between the library staff and the students. The staff had some authority issues as they were obliged to maintain silence in the library although they themselves often felt the mobile phone was not that disturbing. The students found other things much more disturbing than the mobile phone. Accordingly, a whole variety of more or less disturbing noises were mentioned: security alerts, copy machines, computer keyboards, locker doors and especially face-to-face conversations. In most cases the mobile phone was only mentioned after the interviewers had raised this topic explicitly. When questioned about the usage of the mobile phone it clearly turned out that this device had long become a ubiquitous companion for the vast majority of this group of interviewees and it is even used in places where it is usually seen as a disturbance. In this research of disruptions within the library we realized that the mobile phone is but one of many factors that lead to disturbances, highlighting that the mobile phone has lost its exceptional role in this particular context.

Based on this background and the assumption of a certain normalisation of mobile phone usage the idea of a study was born to look not at contexts with 'heavy normative expectations' (Ling, 2004, p. 125), as in the case of the research done in a library, but at the public sphere where the mobile is not only tolerated but even accepted. This new study aimed to look at distinctive social arrangements while using the mobile phone, its intention being to look at peoples' emotional reactions in the environment of mobile phone users after (intentionally) disturbing ringing attacks. The question was not only about whether how people around the recipient could manage the situation but also about a certain threshold value beyond which people would become stressed, troubled and act with sanctioning behaviour. The study is qualitatively oriented and rather than to follow the route of testing pre-formulated hypotheses it follows the idea of a qualitative research strategy where a more open ended approach to methods is an integral part.

## Aim and Methodology of the Study

The study and its findings I present here are part of a broader project that explores the mobile phone and changing communicative practices of everyday life.[1] This sub-study is based on observations in public places complemented by other methods such as personal interviews, group discussions and a paper and pencil questionnaire. Basically this study was an observational study (Lee, 2000), and as such it was the continuation part of a series of non participant observations previously carried out in Italy regarding the mobile phone use on a piazza (Höflich, 2005b, 2006). The empirical work – observations and 'a little bit more' – was carried out from June to September 2007 in Erfurt, Germany. The research was carried out in a street café and in a beer garden in the centre of the city of Erfurt. The street café was open to the pedestrian area where guests of the café could, as a kind of audience, watch the scene around them – but also could be watched by the flâneurs that go past. The beer garden was much more closed, located in an inner courtyard and thus not influenced by passing pedestrians.

The research was divided into two parts. Part one looked at the special arrangements between the users of mobile phones and the third persons present from the point of view of a non- participant observer. In this case third persons sitting on the same table as the user of the mobile phone

1    The study is financed by the German Research Foundation (Deutsche Forschungsge-meinschaft). In particular this sub-study was conducted together with a group of students in the context of a lecture titled: 'The mobile phone in everyday life'. I thank the participants of the seminar for their participation in and contribution to this research.

were of special interest: What is their relation to the present user – may be as a caller or a writer/receiver of mobile written messages? What is their actual behaviour during the mobile phone usage situation? Are they integrated or separated in the context of the usage situation? The second part looks at moments of disturbance and associated reactions – and this part of the research is of particular interest to this paper. At the beginning of the field research the research question was postulated: How do people react in a special case of disturbance via a mobile phone? Do they tolerate or ignore it or are there certain social sanctions that are recognized? The research idea of this second part was especially stimulated by the so called 'breaching experiments' which have been practised in the context of ethno-methodological research by Harold Garfinkel.

The process of producing normality is at the centre of ethnometh-odological research. It looks at the structures of every day life and what is done to produce it. This is actually different from classical sociological thinking. Whilst mainstream sociology is interested in explaining social facts, ethnomethodology aims towards an explication of their constitution (Have, 2004, p. 14). As Kenneth Leiter formulates: 'Ethnomethodology is simply the study of the methods people use to generate and maintain their experience of the social world as a factual object' (Leiter, 1980, p. 25).

More than other theoretical programmes in the field of social science ethnomethodology is best characterized by its research orientation (see Weingarten and Sacks, 1976, p. 7). People take most things for granted in their world and also the procedures that produce it, unless some particu-lar events or eruptions evoke attention. On this latter point I now turn to Garfinkel who had the idea of making trouble within particular environ-ments. As he writes:

> Procedurally it is my preference to start with familiar scenes and ask what can be done to make trouble. The operations that one would have to perform in order to multiply the senseless features of perceived environments; to produce and sustain bewilderment, consternation; to produce the socially structured affects of anxiety, shame, guilt, and indignation; and to produce disorganised interaction should tell us something about how the structures of everyday activities are ordinarily and rou-tinely produced and maintained (Garfinkel, 1967, pp. 37–8).

For instance students were instructed to engage an acquaintance or friend in an ordinary communication situation and then to insist on clarifying a common-sense remark that would otherwise be unnoticed. One example of such a trouble-making procedure looks like this:

> The victim waved his hand cheerily.
> (S) How are you?
> (E) How am I in regard to what? My health, my finances, my school work, my peace of mind, my ...?
> (S) (Red in the face and suddenly out of control) Look! I was just trying to be polite. Frankly; I don't give a damn how you are (Garfinkel, 1967, p. 44).

Such operations 'to produce and sustain anomic features' (Garfinkel, 1990, p. 187) showed (beyond this example) that people confronted with such troubles will initially try to make sense of the eruptions and deviations. By using other words, they will then try to produce some form of normality and so they make 'normal' explanations to make the behaviour of the disturber understandable to themselves:

> If interacting parties sense that ambiguity exists over what is real and that their interaction is thus difficult, they will emit gestures to tell each other to return to what is 'normal' in their contextual situation. Actors are presumed to hold a vision of a 'normal' form for situations, or to be motivated to create one; and hence much of their action is designed to reach this form (Turner, 1987, p. 412).

Thus we learn from Turner and others (see also Joas and Knöbl, 2004, p. 236), that there is even a kind of constraint by the interacting parties to a reciprocal reference so that the behaviour can be seen as meaningful and understandable. In addition, such disturbances might temporarily produce emotional stress – anxiety, shame, guilt and indignation – as Garfinkel asserted. In other words, emotions arise when people are not able to meet the need for 'facticity' (Turner, 2007, p. 123). This is not only true in the case of the person who is confronted with the crisis but also in the case of the troublemaker, because it is also not always easy to break the rules. Garfinkel's experimenters do not say very much about this. But in natural settings people do appear to make reciprocal indications of a common factual world.

In exploring this approach to my study I also considered the scientific status of this kind of crisis experiment. In particular such breach observations may make one sensitive to noticing what is going on (as it will be shown later). As Have states, such troublemaking procedures can be seen as '[...] explicative devices of pedagogical tricks clarifying and demonstrating conceptual issues, rather than research projects as ordinarily perceived' (Have, 2004, p. 41). Indeed, one may ask whether they may even be called experiments. Not least Garfinkel (1967, p. 38) himself emphasized that his studies 'are not speaking experimental' – they are 'aids to sluggish imaginations' that produce 'reflections through which the strangeness of an obstinately familiar world can be detected'. Some scholars express perplexity in respect to this kind of experiment. Gouldner (1974, p. 471) has a rather negative impression and in particular compared these so called experiments with happenings where objectivity and sadism are eloquently combined. Other scholars argue that these experiments lack methodical rules (Kleining, 1986, p. 733). On the contrary, in my study the trouble-making procedures are regarded as an example of a qualitative experiment that Kleining (1986, p. 724) defines as intervention into a social object whose structure is analysed on the basis of scientific rules. The trouble-making procedure is considered as an explorative and heuristic form of the experiment. This intervention makes the qualitative experiment different from observation that is based on reception. Naturally, this intervention inevitably provokes questions of research ethics (where the people should not simply be victims, as Garfinkel calls them). However, based on this specification the breaching experiments can be seen as a kind of premature qualitative experiment because they are more for demonstrating what is already known. Motivated by the breaching experiments this study has a heuristic intention, although they have the similar simple structure as the Garfinkel experiments.

## The Erfurt Study

The approach to my research was to look at the mobile phone usage from the perspective of the productions of a crisis. The ringing of the mobile phone in general was already a real crisis experiment because it – more or less – disturbed any ongoing interaction and because a (partly strange) sound intrudes into the social scenery. This observational study aimed to look at the behaviour of the users of the mobile phones and the reactions of those in their social surroundings and associated social arrangements. But beyond this the idea was to provoke the situational frame and to evoke a question regarding what was actually going on.[2] In this context the reactions of the social environment were observed in addition to the possible processes leading to a normalisation of the situation. The idea of such a kind of experiment in the field of mobile phone research is not entirely new. Rich Ling (Ling, 2002, p. 61) reports (although not with reference to the ethnomethodological breaching experiment) that he did a 'good-natured experiment' to tease out people's reactions to the mobile phone usage in the public sphere. One 'experiment' was to invade the personal sphere of persons who made calls but without the intention to eavesdrop. For instance this was done in a store where he examined the wares immediately beside the caller. Another experiment was to catch the eye of people while making a mobile phone call. He simply looked into their faces as he passed them on the pavement. One result was that individuals yielded the space to the intruder:

> They wandered off to another less populated area. If I were to repeat the experiment with the same subject they would again wander off .... Unlike the metaphor of the boorish ego maniac loudly talking on his [sic] phone, this experiment seems to point

2    The question about 'What is going on here?' just describes what Erving Goffman calls a frame. He writes: 'Whether asked explicitly, as in times of confusion and doubt, or tacitly, during occasions of usual certitude, the question is put and the answer to it is presumed by the way the individuals then proceed to get on with their affairs at hand' (Goffman, 1974, p. 8).

that mobile telephones users are aware of the need to maintain the space around them if possible (Ling, 2002, p. 76).

They felt that a rule of distance was violated and, without sanctioning the intruder, they tried to re-establish a 'normal' distance.

The crisis experiment for this study took place, as mentioned, in a street café and a beer garden in Erfurt, Germany. In both locations a student sat with their mobile phone on his/her table. Two other students were also located nearby to make the calls and to observe the reactions around them. In both instances there were other people, third persons, present sat at nearby tables.

In addition to the experimenter with the mobile phone being observed by his research colleague, they both observed the kind of reactions around each other so that there could be a triangulation of observations. This offered the possibility of observing the observer. The approach to the experiment was to make a call every five to ten minutes but the called student was explicitly not allowed to answer the call. The mobile phone was left to ring again and again with variations of ringing tones. The observer as well as the experimenter wrote down the reactions they perceived. Qualitative research should of course be flexible and open and in this sense the methods had to be extended beyond those originally intended. In this case the methodology of self observation was used to reflect the inner states and feelings of the persons who received the phone calls (see Krotz, 1999). The reasons for this will be shown in the next section. The persons called were instructed to write a detailed account of their inner feelings during the time of the experiments. In addition, after the end of the experimental phase there was a group discussion where the self observations in particular were reflected upon according to the idea of a 'dialogic introspection' (see Kleining, 1999). In sum, there were (besides the nearly one hundred observations in natural settings) nine experimental phases each one lasting about one hour.

## Feelings of Perceived Reactions of the Public and the Emotional Stress of the Mobile Phone User

The observations showed that the ringing of the mobile phone is obtrusive and, as it will be shown, intrusive as well. Indeed, there had been reactions after the ringing as well as during it, but this was not in general. Sometimes people looked towards the person who had been called and was not answering the phone. Sometimes there had been a shake of the head as if perhaps in disapproval or disbelief, but beyond this there have been no other sanctioning behaviours of the third persons around. Just the opposite: in most cases there have been no recognisable reactions. In this case there was a prevalence of civil attention. Another interesting point which became more apparent was that there had been a different perception by the observers and the experimenters. There were also instances where the observers perceived negative reactions and the experimenter could not find any indications of this. But also the opposite was found: the experimenter imagined that people stared at her or him or seemed to speak about her or him. There was a feeling of being regarded as deviant by the experimenter but that was not verified by the external observers. What this tells us is that one needs to be careful with observations and to validate them by a kind of double observation. And above all the differences in observations draw attention to the distinctive situational integration of the acting person. People in a certain situation may see the world around them differently from distant observers. But it is their definition of the situation that has meaning for their action in the sense that if people define a situation as real, it is real in its consequences (Thomas and Thomas, 1973, p. 334). This became increasingly relevant in the context of this study and especially regarding the feeling of stress in a situation of being called on your mobile phone in the public sphere. Above all, in cases where the student experimenter thought that they – or more exactly the ringing of their mobile phone – produced no annoyance they also felt much freer to manage the situation and they even felt some fun in this kind of provocative situation. They did not fear any negative reactions and could regard the situation as a sort of (stimulating) game. However, other students felt just the opposite:

they had the feeling of stress and even fear (of negative sanctions), because they assumed that to provoke attention would instigate negative reactions. Examining this latter point a little more closely we learn from the aforementioned international study that, with only small differences in our data regarding different countries, there were indeed some people, although not the majority – that felt uncomfortable making a call on their mobile phone if strangers were around them (Höflich, 2003).

In the first phase of the Erfurt experiment we found that the so called experimenter (the recipient of the rings) was not able to fulfil the requirements of the study. Although told to let the phone ring and to ignore the phone, all the students took up the mobile phone and looked at it. They even recounted strong feelings of stress and nausea. After this first experimental trial the things that happened were discussed amongst the research team and as a consequence the empirical procedures were adapted. In each of the following experimental phases it was agreed that each student should make an extensive report about his/her inner states, fears and anticipation of the perception of others. On the basis of their reports we learned that throughout the subsequent experiments the students were anything but relaxed. Instead they reported (more or less) feelings of stress and in a few cases even fear, indicating a certain loss of control:

> You know, I felt my pulse, it got higher and higher ...
> My heart is beating. I do not risk raising my head and because of this I can not perceive any reactions around me.
> And then it happened. My mobile was ringing – and the sound of my mobile is not really very unobtrusive. I take my mobile for a short time and look at its display. Anna is calling. I put the mobile back on the table and I try to not look that nervous. I definitely look back. Instead of this I'm flicking through the newspaper. Finally done. The mobile stopped ringing ...

Such feelings are reported as existing before, as well as during the experiment. In particular they reported feelings about being observed and about making a negative impression. One student mentioned:

> Already after I took my place I had the feeling of being totally observed. What should the other guests say? Surely they ask themselves why I'm sitting alone at that table. To be honest I normally don't do this and I also feel uneasy.

There was also a feeling of tension until the next ringing, accompanied with the impression that other people nearby are talking about oneself: 'I have the feeling that the man behind me is something saying like "What is going on" or something like that', and this was accompanied by a fear that there will be an explicit sanction (for instance that they get approached by someone): 'If I know someone is coming, what shall I say?'.

Why all this anxiety? It seemed that all the students mentioned did not want to make any negative impressions. It was not their intention to disturb others – to disturb the normalities of the situation and to break the rules or to evoke awkwardness. These rules for instance indicated, the students presumed, that the mobile phone should not ring very loud, that one had to be considerate about the type of ringing tones, that it should not ring very often and, at very least one had to answer the call. All this indicated that social practices with regard to the appropriate behaviours for mobile phone use in these settings were emerging. In addition it was mentioned by some that there could be a gender effect (but only one of the students participating in the experiment was male!).[3] Others imagined that some people would think that women were not able to turn the mobile onto a silent mode.

In contrast to the Garfinkel experiments it was not the other third parties present who were the victims but the experimenter (the one who got the calls). They had the experience and feelings of what it meant to produce some kind of situational disorder (even if was just imagined). But how did the persons called manage this feeling of being under stress?

A first strategy to cope with the stressful situation could be called hiding. Stress mobilizes your energies for an escape reaction (Ulich and Mayring, 1992, p. 178), but if you cannot escape you should hide. In this sense one student that got a call tried to hide behind his sunglasses:

> The sun is shining and I have my sunglasses on. This is pleasing because I do not have to look directly in the eyes of other people and also I can observe the others

3    Another gender effect is associated with a certain fear of women in the public space. The students involved in the research stress that woman usually do not go into a café or a beer garden without company (see for instance: Ruhne, 2003).

to remain unnoticed. OK, the last was not so important for me at this time. Maybe
after the experiment, if I should be banned from the restaurant and if other guests
should complain about me I will be happy to have those sunglasses to get a little
bit of shelter.

In most cases they avoided visual contact with the persons around
them. Some sought, if possible, some kind of shelter behind other sounds,
for instance a street musician playing with his saxophone was a welcome
acoustic background when receiving calls as it muffled the sound of the
ring tone. 'It wasn't that hard for me because a saxophone musician was
playing loud … I could have done with it much longer …'. Also other 'sounds'
have been welcomed that detract the attention from the ringing of the
mobile phone.

> At the table next to me there was a family sitting with two babies that cried alter-
> natively. Because they have to look after their babies they didn't hear the ringing of
> the mobile. Sometimes crying children have something positive.

In general making calls and accepting the unanswered ringing of the
mobile phone was easier for these researchers if it did not dominate and
draw on the background noise, or in the words of Goffman (1971, p. 46)
if it does not occupy too much 'sound space' (see also Bull, 2004, 2007 or
Kopomaa, 2000). Furthermore, it could also be seen that if the disruptive
situation was in some way shielded from others (such as if the person who
had been called was not alone or if there was some noise around him or
her) the person even felt some pleasure to play the game of 'let the mobile
phone ring'.

Another way of managing the situation was to 'demonstrate' to the
people around that one had some reason or other not to answer or that one
otherwise was not able to react to the call (because one is engaged in an
involving activity that excluded the mobile phone). Usually such kinds of
explanations are nonverbal – using gestures or artefacts. One way is to demonstrate that one intentionally does not answer the call although the ringing has been recognized. This is what normally is expected from you:

> Okay, at least a short look. That one has to do – because one is not deaf. And intui-
> tively I looked at the ringing mobile. To look and to take it back on the table. So the

people around me got to know that I'm not deaf or so. After a while I was trained to ignore it and not to look at it. But at the beginning it was quite hard ... And the thick sun glasses [laughing] ...!

One student for instance tried to signal that a certain person was calling, for instance a former boyfriend, but she explicitly refused to answer. Others pretended to be engaged in another activity– for instance in reading a newspaper or magazine, in writing in a diary or in conversation with a person: 'I'm going on to read [a magazine] as if I didn't hear any tone. I demonstratively write in my diary.' This did not always seem to be really convincing for the so called experimenter:

> I thought that with the *Cosmopolitan* in my hand I'm not so much credible ... After such a long time it was much more difficult to ignore the mobile that so obviously was near me on the table. Additionally, I have already read the page of the magazine. But I did not want to turn over the page. It would feel strange to awake from my stiffness and nevertheless not answer the call. Therefore I still look at the page of my magazine, but I feel a little bit uncomfortable.

What can be seen is that eruptions lead to efforts to restore normality and people 'vigorously sought to make the strange actions intelligible and to restore the situation to normal appearances' (Garfinkel, 1967, p. 47). The persons called almost felt a kind of constraint that demanded of them a clarification of normality – to give the other the impression that they had everything under control. They suggested a meaning that others could take as a motive to act or not to act. Following Goffman one can say 'that the individual constantly acts to provide information that he is of sound character and reasonable competency. When, for whatever reason, the scene around him ceases to provide this information about him, he is likely to feel compelled to act to control the undesired impression of himself he may have made' (Goffman, 1971, p. 163). The persons who had been called were engaged in what Goffman calls remedial work that has the function 'to change the meaning that otherwise might be given to an act, transforming what could be seen as offensive into what can be seen as acceptable' (Goffman, 1971, p. 109). People do this with a 'body gloss', as a relatively self-conscious gesticulation is performed with the whole body. Especially obvious was what Goffman (1971, p. 130) calls 'orientation gloss'. Such a gloss indicates that its provider is oriented in the situation and recognizes what is going on. But also it makes it easier for the others in the gathering to orient towards him. This is, as Goffman emphasizes, a generally important element in social interaction – and also in the situational frame of mobile phone usage.

## Conclusions

The mobile phone is part of everyday life, and also diversely emotionally embedded in the public sphere. There is a reference to public emotions – the usage of the mobile phone in special emotional situations, for instance in dangerous situations, in the context of an assassination or in situations with positive emotions like a soccer game. There is also a reference to emotions in the public, not least because using the mobile phone enables one to be private in the public sphere and also to show an emotional side that normally is part of the private sphere (and not all emotions should be shown in public). 'The very boundary between the public and private is marked in our own and many societies by informal rules about just what emotional performances are permitted in each zone' (Perri 6 et al., 2007, pp. 2–3).

But there was also, as our study has shown emotional stress involved in the public usage of mobile phones. The ringing of the mobile phone confuses the interaction setting: from the perspective of the person being called it is an intrusion, from the perspective of the people around an obtrusion that focuses their attention onto the called person. This can be seen as a version of a contact ability dilemma, which means that everyone wants to reach others but everyone does not, in a reciprocal sense, want to be reached everywhere and every time. Because the person potentially being called has no control this may cause them stress. This becomes particularly obvious if the mobile phone is ringing. But this stress cannot simply be reduced to claims that it is an environmental influence – although indeed it does depend on the concrete contextual features pertaining at the time. The emergence of stress using the mobile phone depends not least on the individual definition of the situation and is a result of transactional relation between the person and the environment (Ulich and Mayring, 1992, p. 178). As the study shows there is not only a feeling of stress as a consequence of a ubiquitous medium, there are also strategies to cope with this. Just because there are new social arrangements emerging regarding a proper use of the mobile phone in the public sphere there is also a constraint not to ignore these rules. In this sense the mobile phone usage is only a further confrontation with the necessity to produce normality in temporally eruptive situations so that the life can go on.

# References

6, P., Squire, C., Treacher, A. and Radstone, S., 'Introduction', in 6, P., Radstone, S., Squire, C. and Treacher, A. (eds), *Public Emotions*, Houndmills: Palgrave, 2007.

Bull, M., 'Automobility and the Power of Sound', *Theory Culture Society* (21), 2004: 243–59.

——, *Sound Moves. iPod Culture and Urban Experience*, London and New York, NY: Routledge, 2007.

Burkart, G., 'Mobile Kommunikation. Zur Kulturbedeutung des "Handy"', *Soziale Welt* (51), 200: 209–32.

Fortunati, L., 'The Ambiguous Image of the Mobile Phone', in Haddon, L. (ed.), *Communications on the Move: The Experience of Mobile Telephony in the 1990s*, Telia: Stockholm, 1998.

——, 'Der menschliche Körper, Mode und Mobiltelefone', *Mobile*, 2005.

Fortunati, L. and Manganelli, A. M., 'La comunicazione tecnologica: Comportamenti, opinioni ed emozioni degli Europei', in Fortunati, L. (ed.), *Telecomunicando in Europa*, Milan: Angeli, 1998.

Garfinkel, H., *Studies in Ethnomethodology*, Cambridge: Polity, 1967.

——, 'Conception of, and Experiments with, "Trust" as a Condition of Stable Concerted Actions', in Coulter, J. (ed.), *Ethnomethodological Sociology*, Aldershot: Elgar Reference Collection, 1990.

Godard, D., 'Same Setting, different Norms: Phone Call Beginnings in France and the United States', *Language and Society* (6), 1997: 209–19.

Goffman, E., *Relations in Public. Microstudies of the Public Order*, London: Allen Lane, 1971.

——, *Frame Analysis: An Essay on the Organisation of Experience*, New York, NY: Harper & Row, 1971.

Gouldner, A. W., *Die westliche Soziologie in der Krise. Vol. 2*, Reinbek bei Hamburg: Rowohl, 1974.

Have, P. ten, *Understanding Qualitative Research and Ethnomethodology*, London, Thousand Oaks, CA, New Delhi: Sage, 2004.

Höflich, J. R., 'Nähe und Distanz. Mobile Kommunikation und das situative Arrangement des Kommunikationsverhaltens', in Grimm, P. and Capurro, R. (eds), *Tugenden der Medienkultur*, Stuttgart: Franz Steiner Verlag, 2005a.

——, 'A Certain Sense of Place. Mobile Communication and Local Orientation', in Nyíri, K. (ed.), *A Sense of Place. The Global and the Local in Mobile Communication*, Vienna: Passagen Verlag, 2005b.

——, 'The Mobile Phone and the Dynamic between Private and Public Communication: Results of an International Exploratory Study', in Glotz, P., Bertschi, S. and Locke, C. (eds), *Thumb Culture. The Meaning of Mobile Phones for Society*, Bielefeld: transcript, 2005c.

——, 'Places of Life – Places of Communication: Observations of Mobile Phone Usage in Public Places', in Höflich, J. R. and Hartmann, M. (eds), *Mobile Communication in Everyday Life: Ethnographic Views, Observations and Reflections*, Berlin: Frank and Timme, 2006.

Höflich, J. R. and Gebhardt, J. (eds), *Kommunikation. Perspektiven und Forschungsfelder*, Frankfurt am Main: Peter Lang, 2005.

Höflich, J. R., Rössler, P. and Gebhardt, J., *Das Handy als Störfaktor in der Universitätsbibliothek Erfurt*, unpublished research report, University of Erfurt, 2007.

Joas, H. and Knöbl, W., *Sozialtheorie. Zwanzig einführende Vorlesungen*, Frankfurt am Main: Suhrkamp, 2004.

Kleining, G., 'Das qualitative Experiment'. *Kölner Zeitschrift Soziologie und Sozialpsychologie* (38), 1986: 724–50.

——, 'Vorschlag zur Neubestimmung: Dialogische Introspektion', *Journal für Psychologie* (7), 1999: 17–19.

Kopomaa, T., *The City in Your Pocket. Birth of the Mobile Information Society*, Helsinki: Gaudeamus, 2000.

Krotz, F., 'Forschungs- und Anwendungsfelder der Selbstbeobachtung', *Journal für Psychologie* (7), 1999: 9–11.

Lee, R. M., *Unobstrusive Methods in Social Research*, Buckingham: Open University Press, 2000.

Leiter, K., *A Primer on Ethnomethodology*. New York, NY and Oxford: Oxford University Press, 1980.

Ling R., *The Mobile Connection. The Cell Phone's Impact on Society*, Amsterdam: Morgan Kaufmann, 2004.

——, 'The Social Juxtaposition of Mobile Telephone Conversations and Public Spaces', *The Social and Cultural Impact/Meaning of Mobile Communication: Chunchon Conference on Mobile Communication*, 13–14 July 2002.

Mullett, M. M., 'How we Behave when we Telephone', *American Magazine* (86), 1918: 44–5, 95.

Ruhne, R. R., *Macht, Geschlecht. Zur Soziologie eines Wirkungsgefüges am Beispiel von (Un)Sicherheiten im öffentlichen Raum*, Opladen: Leske und Budrich, 2003.

Thomas, W. I. and Thomas, D. S., 'Die Definition der Situation', in Steinert, H. (ed.), *Symbolische Interaktion. Arbeiten zu einer reflexiven Soziologie*, Stuttgart: Klett, 1973.

Turner, J. H., *The Structure of Sociological Theory*, revised edition, Homewood, IL: The Dorsey Press, 1978.

——, *Human Emotions. A Sociological Theory*, London and New York, NY: Routledge, 2007.

Ulich, D. and Mayring, P., *Psychologie der Emotionen*, Stuttgart, Berlin, Köln: Kohlhammer, 1992.

Weingarten, E. and Sack, F., 'Ethnomethodologie. Die methodische Konstruktion der Realität', in Weingarten, E., Sack, F. and Schenkein, J. (eds), *Beiträge zu einer Soziologie des Alltagshandelns*, Frankfurt am Main: Suhrkamp, 1976.

# Decorated Mobile Phones and Emotional Attachment for Japanese Youths

SATOMI SUGIYAMA

## Introduction

Scholars who examine the mobile phone from the perspectives of communication, sociology, and anthropology, have identified and characterized symbolic aspects of the mobile phone (e.g. Katz, 1999, 2003; Katz and Aakhus, 2002; Katz and Sugiyama, 2005, 2006; Campbell, 2002; Campbell and Russo, 2003; Fortunati, 2002a, 2003; Fortunati et al., 2003; Green, 2003; Ling and Yttri, 2002; Ling, 2003; Oksman and Rautiainen, 2003). A significance of such findings in considering interpersonal communication is that they suggest that the mobile phone is conceived as an object for expressing the self. Although other objects can serve the process of self-expression, the mobile phone is a unique self-expressive object because people tend to carry it regularly for 'perpetual contact' (Katz and Aakhus, 2002) with friends and family members.

Katz and Aakhus state that just as machines are in 'perpetual motion', people with personal communication technologies are in 'perpetual contact'. The theory of Apparatgeist (Katz and Aakhus, 2002), as well as the notion of perpetual contact, provides a theoretical framework that attempts to explain the social process focusing on technologically mediated interpersonal communication. Katz and Aakhus state:

> Social actors must constantly perform a series of ever-changing and highly complex social roles. They must also deal with other actors who themselves are performing a series of ever-changing and highly complex social roles. This in itself represents an uncertain and complex scenario in which communication takes place (Katz and Aakhus, 2002, p. 314).

The already complex and dynamic social interaction is further complicated by the presence of the mobile phone, suggesting that it is important
to consider the sense of self and its expression in the social interactions
involving mobile communication technologies.

It is within this context that emotion becomes relevant to the mobile
phone research. Vincent (2003) and Fortunati (2005) have noted the significance of numerous emotions that the mobile users experience. Vincent
argues that users of the mobile phone show emotional attachment not to
the device itself but to 'the content it enables, the relationships it maintains and the information stored on it' (Vincent, 2003, p. 224). Fortunati
points out that not only users but also bystanders tend to have various
emotional reactions.

Japan has been gaining much attention due to its unique technological development of the mobile phone (e.g. Ito et al., 2005; the success of
i-mode, see Barry and Yu, 2002; Yu and Tng, 2003; the widespread use of
webphone, see Miyata et al., 2005a; Miyata et al., 2005b). As a country
that can be characterized as 'the epicenter of mobile communications'
(Srivastava, 2008), Japan serves as a cultural context worth examining.
Furthermore, as Miyata, Boase, and Wellman (2008) state, young people
in Japan who 'nurture relationships that might otherwise be hampered by
parents and other authority figures' (Miyata et al., 2008, p. 209), form a
core of mobile users. In particular, college is the life-stage in which many
teenagers start spending significantly more time with their own peers, and
the kind of people to which they relate extends beyond classmates, diversi-
fying their social network in terms of age, upbringing, social status, and so
on. They develop and maintain such social networks while they either live
with their parents or often alone in an apartment since most universities
in Japan do not offer residential colleges. Then, how do college students in
Japan experience emotions in relation to their mobile phone as they negotiate various interpersonal relationships in their daily social interactions?

This chapter examines the emotions that Japanese youths experience as
they use their mobiles, particularly focusing on college students. The data
to be presented here were collected and analyzed in pursuit of answering
the large research question of symbolic meanings of the mobile phone for
Japanese youths. What emerged in the process was the realization that the

mobile phone has become a locus for collecting personal and interpersonal significations that express and define the self in public places, and that these young people in Japan develop emotional attachment to their mobile phones. In this chapter, I build upon the aforementioned finding by Vincent (2003), and seek to answer the above research question by delineating how youths in Japan do develop emotional attachment to stored content and relationships, but also grow attached to the instruments themselves.

## Study Participants

Four sessions of focus group interviews were conducted in Japan early in 2005. Japanese college students were recruited for the study at a private university in Nagoya. The interview participants were considered as appropriate in that they have 'appropriate experience in the cultural scene' (youth culture in this case), and 'the ability and willingness to articulate his or her experience in the interview context' (Lindlof, 1995, p. 178). Thirty students participated. The first two groups consisted of first year students (total of 17 students) and another two groups (total of 13 students) were upper year students. Since Japanese communication style is often characterized as hierarchical and likely to be sensitive to age differences when expressing opinions, younger students might refrain from expressing their honest opinions, thus creating a superficial conformity among the group members. Therefore, the groups were formed in such a way that first year students did not mix with upper year students. This follows the recommendation by Lindlof, that a focus group interview be composed of 6–12 people: '[...] who are demographically homogeneous, or who have certain experiences in common' (Lindlof, 1995, p. 174).

Among the first year students, 3 male students and 14 female students participated. More specifically, one group of the first year students was composed of 8 female students, and another group was composed of 3 male students and 6 female students. Among the upper year students, 3 male students and 10 female students participated. More specifically, one group of the upper year students was composed of 5 female students and

2 male students, and another group was composed of 5 female students
and 1 male student.

## Procedure

Each session took place in a quiet university classroom. Since the classroom offered a familiar environment for the participants, the setting
helped to create a comfortable atmosphere. The semi-directive approach
was employed so that the interview participants discussed the topic of
concern, while maintaining maximum possible freedom in the direction
of the conversation. In order to build good rapport with the interview
participants, I made sure to create a casual and informal atmosphere using
informal conversation styles. Since the interviewer tends to serve 'as a co-
participant in the construction of a discourse' (Briggs, 1986, p. 25), I sought
to function as a moderator of the conversation, attempting to minimize
leading the discussion into any particular direction. Each session lasted one
to one and a half hours. All interviews were audio-recorded.

## Data Analysis

Once the interviews were completed, I transcribed all recorded interviews
(unabridged transcripts). This procedure yielded 103 pages of interview text
to be analyzed (in Japanese language). Then, I read the interview transcripts
several times, recommended by the initial step of a qualitative analysis
(Maxwell, 1996). As a second step, I examined some broad themes in the
interview transcripts. In this process, I generally followed a coding tech-
nique by Strauss and Corbin (1990). Throughout the process of coding,
constant comparative analysis (Strauss and Corbin, 1990) was employed.
Once the analysis was completed, the interview quotes that significantly
illustrate the themes and discussion points were translated into English
and were incorporated into the presentation of the data analysis.

# Emotional Attachment to the Relationship that the Mobile Maintains

The focus group participants discussed their mobile phone experience using some emotional terms. This point becomes evident in a relational context. A notable point that emerges from the data is that these youths in Japan see the mobile as an object that offers perpetual contact with their relationally significant others. For example, the following interview quotes indicate that they rely upon the potential interactions that the mobile can offer to the point that they need the 'courage' to turn it off.

> Courage to turn off the mobile phone
> (Focus group #4, line 949–55, upper year females)
> FEMALE 1: I don't have the courage to turn off my mobile phone.
> MODERATOR: You need the courage to turn it off?
> Several: Yeah.
> FEMALE 1: Well, somehow, the power button [of the mobile phone] ... wonder how to say ... if I turn off the phone, if I receive a phone call, it won't be recorded as a received call ... if it's a mail, it will reach [me], but I cannot figure out what happened when, but if I keep it in 'manner' mode[1], I can know that I received a phone call, and I can connect [to the person] right away. As much as possible, I avoid turning off my phone.
> MODERATOR: Do you turn off your mobile phone or not?
> FEMALE 2: No. I almost lost my friend.
> MODERATOR: So you don't turn it off.
> FEMALE 2: I can't [turn it off]. Well, in the past, on my birthday ... [others start giggling] you guys were so helpful at the time! A mail message did not reach me, and on the day, the person with whom I was planning to hang out, the mail message did not reach me, and I was misunderstood for not having contacted the person. Ever since, our relationship became awkward.

The comment by female 1 above shows her reason for requiring courage to turn off the phone, revealing the anxiousness of not knowing, however briefly, who tried to reach her. The comment by female 2 above saying that

---

1   'Silent' mode in the United States is commonly called 'manner' mode in Japan.

she 'can't' turn it off, shows how her interpersonal relationships became at risk through an accidental malfunction of the mobile phone. Although she was telling this story rather jokingly, it was quite apparent from the tone of her voice that she took the incident quite seriously. The above quotes indicate how much their interpersonal relationships rely upon the mobile phone. To them, turning off the mobile phone is almost like cutting a social blood vein. This does not only mean a temporary disconnection from their relational partners, but also a disconnection that might have serious relational consequences, causing some blockages in the social vein. For some people, turning off the mobile phone is just a mere action of pushing a button, but to others, it means much more.

The emotional attachment directed toward the potential relational consequences was also expressed in the participants' description of their experience exchanging mails (meiru).[2] Numerous students who participated in the interview shared ambivalent feelings about exchanging mails. They said that they feel quite uncertain about what the text-based messages convey, even if they involve pictographs and emoticons. On this subject, they report several episodes of misunderstanding or being misunderstood as 'being mad'. Some also reported that they feel burdened by being obligated to return a mail once they receive it, even when they are busy. Of course, 'no immediate reply' sends a relational message that is open to numerous interpretations.

The students reported that because they feel emotional attachment to the relationships the mobile can facilitate, they feel comfort/relief when they are *with* their mobile phone. The word *anshin* (comfort/relief) came up in all four groups in the context that they feel 'comfort' when they are with their mobile phone. On the other hand, if they accidentally left it at home, they say that they feel very anxious and uneasy. One female student (FGI 1) even said that she feels like just going home as soon as possible if she left her mobile phone at home.

---

2    'Meiru' (mail) is a typical way that Japanese users refer to 'text messages sent as short messages and those sent as Internet email' (Matsuda, 2005, p. 35).

Gergen (2002) and Fortunati (2002b) pointed out that one of the challenges involved in the way we use the mobile phone is absent presence/present absence. Drawing the concept of absent presence, Vincent (2003) argued that 'the tension between needing the mobile and concern at losing it and all that it contains' becomes intensified by 'the potential loss of the relationships that the mobile delivers either directly or via their "absent presence"' (Vincent, 2003, p. 221). The case of Japanese youths captures this tension, being expressed as a mixture of emotions such as comfort, uneasiness, and anxiousness.

As the above discussion suggests, the mobile phone is certainly, at least partially, a medium that expands our communicative capacity with distant others. It offers us a potential for perpetual contact as well as absent presence. However, the mobile appears to be evolving into something more. One of the points that emerged from the data is that these focus group participants feel that the mobile phone has a significant presence in their life. During a focus group interview, the following conversation took place:

> (Focus group #1, line 87–92, first year females)
> FEMALE 1: I feel uneasy.
> [Other participants show agreement nonverbally]
> MODERATOR: Why do you feel uneasy?
> FEMALE 1: I somehow cannot help checking it.
> [Other participants agree]
> FEMALE 1: Even if I keep it silent ...
> [Other participants agree]
> FEMALE 1: I know it's there, so I can't ignore it and feel like going to check it ... it has such a huge presence.
> OTHER PARTICIPANTS: Yeah, huge, huge! [with laugh]
> FEMALE 2: The reason why I feel uneasy when I left it at home, and I cannot forget about it even when I am studying is that it has really a huge presence [in my life].

Although two particular female students were commenting on their experience, it was a comment to which other participants were clearly able to relate and agree. The mobile phone is something that they cannot ignore, and they feel its prominent presence. The above quote captures a rather complex emotion associated to the mobile phone: the small machine seems to be turning into a monster that has a huge presence.

# Emotional Attachment to the Mobile Itself

These Japanese college students, who would like the mobile phone to be always beside them, reported that they take it everywhere including the toilet and bathroom.[3] Quite naturally, accidents happen. It was quite surprising that so many female students reported that they have dropped their mobile phone in the toilet and bathtub. And as one can imagine, it is a significant event for them. In describing such an event, one student said, 'Isn't it sentimental [as you see off the mobile in the toilet]?'[4] Such an event is not just upsetting for her but it provokes 'sentimental' feeling. Her expression was reminiscent of the moment seeing off a beloved one. It was just that her beloved one, in this case, was her mobile phone, disappearing into the toilet.

Another student, who has experienced 'sunk in the water' accident in the bathroom, now wraps her mobile phone with a towel, and leaves the towel-wrapped phone right outside of the bathroom door so that she can be with her mobile phone the moment she gets out of the bath.

Hard to be separated from the mobile phone
(Focus group #4, lines 704–17, upper year females)
FEMALE 1: I used to carry the mobile phone everywhere. Even to the toilet and bath.
ALL: [laugh]
MODERATOR: You said earlier that you had an experience of dropping it in the bathtub?
FEMALE 1: I learned a lesson [from the experience], and I don't take it to the bath. My friends told me that it's my fault that I dropped it in the bathtub since I took it there.
ALL: [laugh]
FEMALE 2: If I am in the middle of exchanging mail, I feel bad about interrupting the exchange, so I wrap my mobile phone in the towel ...
FEMALE 3: You wrap it in the towel?
FEMALE 1: Yeah!

3   The toilet and bathroom are typically separate rooms in a house or apartment in Japan.
4   For more details see Sugiyama (2006).

FEMALE 2: Yeah, wrap it in the towel like this [with gesture], and take it to the bath.
FEMALE 3: I don't take it to the bath.
FEMALE 2: Well, when I take it to the bath [I wrap it in the towel].
MODERATOR: Do you?
FEMALE 4: Once in a while. But I don't do it very often because I am scared [of dropping it]. But I make sure to leave it right outside of the bath so that I can touch it quickly.

This conversation of the interview participants, who are good friends among themselves, reveals their great resistance for being separated from the mobile. However, it is also an object likely to be ruined under certain condition, in this case, when dropped in water. The tension between the desire to keep the mobile close and the desire to protect it is revealed in a hesitant comment reporting leaving it behind when they take a bath and some compromising solutions that they devised such as wrapping it in the towel or keeping it right outside of the bathroom. As the above conversation shows, not all young people engage in this behaviour. However, those who do count as evidence of the close and intimate relationship that may be developed with the mobile phone. The intimate feeling was also expressed in the way female 2 in the above quote talked about this ritual and the care with which her gestures illustrated how she wrapped her mobile with a towel. It was as if she was talking about something precious. These sentimental feelings, namely intimacy, caring, and resistance from being separated, all appear to signal that the youths in Japan are developing attachment to the mobile itself. And this might happen because they start blurring the lines between the perpetual contact with their relational partners that the mobile can offer and the machine itself. The two are no longer separable and this increases in their experience the significance of the machine itself: something that they feel very close to themselves.

# Internally and Externally Decorated Mobile Phones

How do these young Japanese students come to experience emotional attachment to the machine? There are two aspects that can be delineated from the data: emotional attachment to the mails exchanged and stored in the phone and emotional attachment to the items attached to the surface of the mobile phone.

## *Internally Decorated Mobile Phones*

Participants reported that exchanging mails (*meiru*) is a predominant way of using their mobile phone. In describing this behaviour, some themes that indicate their emotional attachment to the mobile have emerged. For example, an upper year female student stated:

> Personal signification in the mobile phone
> (Focus group #4, line 983, female)
> The mobile phone is where I keep my memories, I think. Both of telephone calls and mail. When I receive very emotionally moving mail, I protect it (*hogo suru*) so that I can save it.

It is notable that she used the term 'protect' (*Hogo suru*) for 'saving' (*Hozon suru*) past mail. This is, in fact, indicative of how significant some of this mail is to her. Some mail that arrives in her little machine can be 'emotionally moving'. It is so precious that she would like to 'protect' it, and that she can reflect upon her 'emotionally moving' experience again and again.

Not only this participant but others, too, reported that they take time to review their received calls and mails when they have a moment. On the train, in particular, many said that they somehow take out the mobile phone as soon as they sit down, and start reading the past mail exchanges with some nostalgic feeling. They cherish the 'protected' messages as well as the more recently exchanged mail by reviewing them later on.

A group of female students reported that they enjoy counting with their friends the number of exchanges that they accumulate. They mentioned that some mobile phones automatically count the number of the exchanges and indicate it after the 'RE' (reply) signs in the title of the message. They reported that they are eager to reach a higher number, feeling happy about accumulating 'RE's. In this way, these young college students collect personal memories into their mobile phone. The mobile phone is, indeed, a place where numerous interpersonal relationships are negotiated. This process of negotiation is recorded in the form of mails, and in the storing of overflowing personal memories in the mobile phone. As a focus group participant noted, the mobile phone is like 'a treasure box of privacy' (focus group #1, first year female). In the same way one opens up a treasure box, these young people in Japan might be opening up their mobile phones that are filled with their sentimental treasures. What the three examples above illustrate is that these Japanese young people are *decorating* the inside of the mobile phone through this interpersonal negotiation. That is, by 'protecting' (*Hogo suru*) certain records of interactions with others, they are internally decorating their mobile phone to their taste. It is a decoration because the stored mail serves as embellishments that shape the inside of the mobile. Text-based mail can be considered as equivalent to photos in the sense that both are embedded with relational significances and memories. In the same way that one decorates the inside of a house with photos with family and friends, they decorate the inside of their mobile phone with relationally meaningful materials. And the internally decorated mobile phone is available for their perusal whenever they need or feel disposed to do so.

## Externally Decorated Mobile Phones

The focus group interviews of Japanese college students showed that they seek to reflect their individuality and taste in mobile phones by selecting a particular design/style as well as by decorating the off-the-shelf devices (Sugiyama, 2006). This resembles the case of Finnish teens that Oksman and Rautiainen (2003) reported. However, the mobile phone is not only a locus for reflecting their individuality. A few female participants had dangling charms of the Hello Kitty character. They put it on because they got it from one of their friends as a souvenir (e.g., Focus group #2, line 492, female; Focus group #2, line 304, female; Focus group #4, line 596, female). Some first year female students explained that it had been popular to give a 'regional Hello Kitty' charm as a souvenir, and once they receive such a gift, they 'feel bad' not putting it on something.

Upper year female students made similar remarks. They said that the mobile phone is the easiest place on which to put the charm. One also reported that she feels happy when she finds that the key chain that she gave to her friend is attached to her friend's mobile phone. Another also agreed, saying that she would probably feel upset if her gift is not on her friend's mobile phone, inattentively being left at home. In addition, quite a few reported that they bought identical dangling accessories when they were with their friends shopping together for their mobile phones. This shows that the accessories that Japanese youths place on their phones seem not only to be a way of showing individuality but also seem to be a way of manifesting social relationships.

Another popular form of decoration for the mobile phone was found to be 'Print Club' (*purikura*) stickers. *Purikura* stickers are photo stickers produced in a small digital photo booth found in numerous locations across Japan. The *purikura* stickers have been extremely popular since 1995, particularly among young girls. They often place the photo stickers taken with their friends in diaries or in special albums (Chalfen and Murui, 2004). The stickers were found to be a widespread form of decoration for the mobile phone among the youths in Japan. Since their inception, *Purikura* stickers have been playing the role of a social currency and work as a manifestation of social connectedness among Japanese youths (Chalfen and Murui, 2004). The mobile phone, which is frequently used in public places, became a perfect place for displaying the social connectedness via *purikura* stickers. Those stickers differentiate one's own mobile phone from others even if the mobile looked exactly the same when it was purchased at a store. Among the 30 focus group participants, at least 9 participants had *purikura* stickers on their mobiles on the day of the focus group interview.

Another example from the focus group reveals an intricate negotiation of the self and social relationships involving mobile phone decoration. Upper year students reported that it is popular to paste a *purikura* sticker of their boyfriend/girlfriend on the inside of the battery cover or on the battery pack itself, so that onlookers cannot see the sticker. This special spot is reserved for a special person, and the *purikura* stickers of their friends are deemed to deserve the more visible places of the mobile phone. One male participant said some of his male friends have a *purikura* sticker on this hidden place, and one female participant said that she actually has a sticker placed on the spot. One female student who has engaged in this behaviour in the past said that she has done it because she was shy about putting a photo with her boyfriend on a visible spot. The hidden photo is shown only to those who the owner of the phone considers to be close friends. Some reported that the conduct was once very trendy among their friends. This exemplified Goffman's notion of public performance (1959).[5] These Japanese young people are carefully managing the way they present themselves to the eyes of the audience of the performance. They seem to think that the photos with their friends are appropriate ways of present-ing themselves to the vast public, but the photos with their boy/girlfriend are not. This suggests that they are quite aware that their mobile phone is exposed to the eyes of countless others. Since the mobile phone serves as a personal front (Fortunati, 2005; Sugiyama, 2006), they need to make the expressive equipment appropriate for the self that they would like to present in public. In this case, the mobile phone serves as a locket, secretly holding a feeling of intimacy and a photo of a beloved one.

## The Decorated Mobile as a Relational Artefact

As the discussions above highlight, this collage of personal significations, which are collected in and on the mobile phone, transforms the phone into a special object for which the owner feels emotional attachment. Even if they do not particularly like the Hello Kitty charm, mobile owners might

5    Richard Ling has been conducting Goffmanian analysis of the mobile phone extensively, including in his recent works (2004, 2008).

put it on their mobile phone to maintain the relationship with the giver of the charm, as well as to maintain their understanding of their self for themselves and for others.

One might also develop an emotional attachment to the charm as the interpersonal relationship with the giver becomes more special. As one negotiates everyday social interactions, the symbolic meanings of each decorative item continually change, contributing to the overall meanings of the mobile phone. And they start developing *aichaku* (emotional attachment). *Aichaku ga waku* means that some emotional attachment starts emerging/growing out of their inner self. These young people in Japan collect contact information of others, photos, and messages inside of their mobile phones, and collect gift charms and *purikura* stickers with their friends and boyfriend/girlfriend on the surface of their mobile phones. This process seems to be not only a way of defining themselves and negotiating their social relationships, but also a way of cherishing their own mobile phones. This can be interpreted as the process of domestication, which involves 'a taming of the wild and a cultivation of the tame' (Silverstone and Haddon, 1996, p. 60).

The mobile is entrusted with significant memories (as diaries or journals once were), becoming dear enough to hold the images, words, and souvenirs of loved ones. Through the act of creating the collage within and on the surface of the mobile phone, some of these young people appear to have embodied their mobile phones, developing emotional attachment to it. It is in this appropriation process that the emotional attachment develops. Silverstone and Haddon state: '[...] in their ownership and in their appropriation into the culture of family or household and into the routines of everyday life, they are at the same time, cultivated' (1996, p. 60).

It is important to broaden the scope of the domestication theory to include space outside the home and the household when examining the mobile phone (Katz and Aakhus, 2002; Haddon, 2003), because the mobile phone is not disappearing from the public sphere as a result of 'domestication', but rather it intrudes even further into our everyday life outside of the home ever more (Katz and Sugiyama, 2006). In the process of domestication and emotional attachment, the mobile becomes a natural part of the body (Fortunati, 2003, 2005), and becomes us in the sense that mobiles become incorporated into our self (Katz, 2003), blurring the line between the mobile and the users.

No matter how much the technology is entrusted and cherished, however, it is also notable how quickly the decoration of mobile phone changes. The interviews indicated that the collage of the mobile phone decoration is constantly changing. Some of the participants came to the interview with a brand-new charm that they just received for free from their school cafeteria. They also place new stickers when they 'get tired of' their mobile phone. If a *purikura* sticker is peeling off, they put on a new one. Some also report that unless they have a 'special memory' associated with the particular charm, they do not carry the accessories over to their new mobile phone device because accessories have to match with the colour and design of the mobile phone device. This constant modification of the mobile phone's appearance suggests that the state of this small object, to which these young people become emotionally attached, is in flux. They continuously modify the internal and external decoration of their mobile, yielding a transitory relational artefact that simultaneously symbolizes their self at the moment.

## Conclusion

The mobile renders itself as a place upon which the significances of life are collected and unified. This happens not only through decorating inside the mobile phone by storing personal information, messages, and photographs, but also as a result of decorating the surface of the mobile phone by attaching charms and photo stickers. Through this continuous process of crafting and creating the internally and externally decorated mobile phone it is understood as a relational artefact, as well as an expressive collage of the self, to which these young people in Japan develop emotional attachment. By cherishing both the inside and the outside of the mobile, the decorator strengthens its emotional connection with the miniature monstrous artefact, even if the connection is short-lived.

Emotional attachment to the relationships that the mobile maintains grows with the potential for perpetual contact and the perceived intimacy that absent presence provokes. Storing exchanged mails and photos, as well as attaching charms and stickers that have personal and relational meanings is the process of attaching themselves emotionally and symbolically. By actualizing this 'monstrous' technology that can make their lives hectic and anxious, the mobile phone becomes a natural part of their everyday life. With the emotional attachment and the increasing proximity between the object and the self, the mobile phone is domesticated into their everyday life as a relational artefact, which simultaneously expresses and defines the self. This echoes the ambivalent identity of the mobile phone that Fortunati (2002b) described. Turkle points out that '[...] the questions raised by relational artefacts are not so much about the machines' capabilities but our vulnerabilities – not about whether the objects *really* have emotion or intelligence but about what they evoke in us' (2007, p. 597).

The mobile phone can be considered as a relational artefact not only in the sense that it enables absent presence of relational partners but also in the sense that it is an object where all relational significances gather. If this is the case, then, the emotions that the mobile phone evokes in us warrant continued examination in order to understand humans in relation to technology.

## References

Barry, M. and Yu, L., 'The Uses and Meaning of I-mode in Japan', in *Revista de Estudios de Juventud* (57), 2002: 139–50.

Briggs, C. L., *Learning How To Ask: A Sociolinguistic Appraisal of the Role of the Interview in Social Science Research*, New York, NY: Cambridge University Press, 1986.

Campbell, S. W., *The Social Construction of Mobile Telephony: An Application of the Social Influence Model to Perceptions and Uses of Mobile Phones within Personal Communication Networks*, doctoral dissertation, University of Kansas, 2002.

Campbell, S. W. and Russo, T. C., 'The Social Construction of Mobile Telephony: An Application of Social Influence Model to Perceptions and Uses of the Mobile Phones within Personal Communication Networks', *Communication Monographs* (70), 2003: 317–34.

Chalfen, R. and Murui, M., 'Print Club Photography in Japan: Framing Social Relationships', in Edwards, E. and Hart, J. (eds), *Photographs Objects Histories: On the Materiality of Images*, New York, NY: Routledge, 2004.

Fortunati, L., 'Italy: Stereotypes, True and False', in Katz, J. E. and Aakhus, M. (eds), *Perpetual Contact Mobile Communication, Private Talk, Public Performance*, Cambridge: Cambridge University Press, 2002a.

——, 'The Mobile Phone: Toward New Categories and Social Relations', *Information, Communication and Society* (5), 2002b.

——, 'The Human Body: Natural and Artificial Technology', in Katz, J. E. (ed.), *Machines That Become Us*, New Brunswick, NJ: Transaction, 2003.

——, 'Mobile Telephone and the Presentation of Self', in Ling, R. and Pedersen, P. E. (eds), *Mobile communications: Re-negotiation of the Social Sphere*, London: Springer, 2005.

Fortunati, L., Katz, J. E. and Riccini, R., 'Conclusion: Bodies Mediating the Future', in Fortunati, L., Katz, J. E. and Riccini, R. (eds), *Mediating the Human Body*, Mahwah, NJ: Lawrence Erlbaum, 2003.

Gergen, K. J., 'The Challenge of Absent Presence', in Katz, J. E. and Aakhus, M. (eds), *Perpetual Contact Mobile Communication, Private Talk, Public Performance*, Cambridge: Cambridge University Press, 2002.

Goffman, E., *The Presentation of Self in Everyday Life*, New York: Anchor, 1959.

Green, N., 'Outwardly Mobile: Young People and Mobile Technologies', in Katz, J. E. (ed.), *Machines That Become Us*, New Brunswick, NJ: Transaction, 2003.

Haddon, L., 'Domestication and Mobile Telephony', in Katz, J. E. (ed.), *Machines That Become Us*, New Brunswick, NJ: Transaction, 2003.

Ito, M., Okabe, D. and Matsuda, M., *Personal, Portable, Pedestrian: Mobile Phones in Japanese Life*, Cambridge, MA: MIT Press, 2005.

Katz, J. E., *Connections: Social and Cultural Studies of the Telephone in American Life*, New Brunswick, NJ: Transaction Publishers, 1999.

—— (ed.), *Machines That Become Us*, New Brunswick, NJ: Transaction, 2003.

Katz, J. E. and Aakhus, M., *Perpetual Contact Mobile Communication, Private Talk, Public Performance*, Cambridge: Cambridge University Press, 2002.

Katz J. E. and Sugiyama, S., 'Mobile Phones as Fashion Statements: The Co-Creation of Mobile Communication's Public Meaning', in Ling, R. and Pedersen, P. E. (eds), *Mobile Communications: Re-negotiation of the Social Sphere*, London: Springer, 2005.

——, 'Mobile Phones as Fashion Statements: Evidence from Student Surveys in the US and Japan', *New Media and Society* (8), 2006.

Lindlof, T. R., *Qualitative Communication Research Methods*, Thousand Oaks, CA: Sage, 1995.

Ling, R., 'Fashion and Vulgarity in the Adoption of the Mobile Telephone among Teens in Norway', in Fortunati, L., Katz, J. E. and Riccini, R. (eds), *Mediating the Human Body*, Mahwah, NJ: Lawrence Erlbaum, 2003.

——, *The Mobile Connection: The Cell Phone's Impact on Society*, San Francisco, CA: Morgan Kaufmann, 2004.

——, *New Tech, New Ties: How Mobile Communication is Reshaping Social Cohesion*, Cambridge, MA: MIT Press, 2008.

Ling, R. and Yttri, B., 'Hyper-Coordination via Mobile Phones in Norway', in Katz, J. E. and Aakhus, M. (eds), *Perpetual Contact Mobile Communication, Private Talk, Public Performance*, Cambridge: Cambridge University Press, 2002.

Matsuda, M., 'Discourses of *Keitai* in Japan', in Ito, M., Okabe, D. and Matsuda, M. (eds), *Personal, Portable, Pedestrian: Mobile Phones in Japanese Life*, Cambridge, MA: MIT Press, 2005.

Maxwell, J. A., *Qualitative Research Design: An Interactive Approach*, Thousand Oaks, CA: Sage, 1996.

Miyata, K., Boase, J. and Wellman, B., 'The Social Effects of Keitai and Personal Computer E-mail in Japan', in Katz, J. E. (ed.), *Handbook of Mobile Communication Studies*, Cambridge, MA: MIT Press, 2008.

Miyata, K., Boase, J., Wellman, B. and Ikeda, K., 'The Mobile-izing Japanese: Connecting to the Internet by PC and Webphone in Yamanashi', in Ito, M., Okabe, D. and Matsuda, M. (eds), *Personal, Portable, Pedestrian: Mobile Phones in Japanese Life*, Cambridge, MA: MIT Press, 2005a.

Miyata, K., Wellman, B. and Boase, J., 'The wired – and wireless – Japanese: Webphones, PCs and Social Networks', in Ling, R. and Pedersen P. E. (eds), *Mobile Communications: Re-negotiation of the Social Sphere*, London: Springer, 2005b.

Oksman, V. and Rautiainen, P., '"Perhaps It Is a Body Part": How the Mobile Phone Became an Organic Part of the Everyday Lives of Finnish Children and Teenagers', in Katz, J. E. (ed.), *Machines That Become Us*, New Brunswick, NJ: Transaction, 2003.

Silverstone, R. and Haddon, L., 'Design and the Domestication of Information and Communication Technologies: Technical Change and Everyday Life', in Mansell, R. and Silverstone, R. (eds), *Communication By Design*, New York: Oxford University Press, 1996.

Srivastava, L., 'The Mobile Makes Its Mark', in Katz, J. E. (ed.), *Handbook of Mobile Communication Studies*, Cambridge, MA: MIT Press, 2008.

Strauss, A. and Corbin, J., *Basics of Qualitative Research: Grounded Theory Procedures and Techniques*, Newbury Park, CA: Sage, 1990.

Sugiyama, S., 'Fashioning the Self: Symbolic Meanings of the Mobile Phone for Youths in Japan', doctoral dissertation, Rutgers, NJ, State University of New Jersey: AAT3240282, 2006.

Turkle, S., 'Authenticity in the Age of Digital Companions', *Interaction Studies* 8 (50), 2007: 1–517.

Vincent, J., 'Emotion and Mobile Phones', in Nyiri, K. (ed.), *Mobile Democracy: Essays on Society, Self and Politics*, Vienna: Passagen Verlag, 2003: pp. 215–24.

Yu, L. and Tng, T. H., 'Culture and Design for Mobile Phones for China', in Katz, J. E. (ed.), *Machines That Become Us*, New Brunswick, NJ: Transaction, 2003: pp. 201–17.

Theme 2

Emotion Mediated through the Contents ICTs Convey

# The Myth of Impoverished Signal: Dispelling the Spoken Language Fallacy for Emoticons in Online Communication

NAOMI BARON

## Introduction

Human communication consists of more than sounds, graphemes, words, and sentences.[1] It also expresses emotions. Sometimes such emotions are conveyed through lexical or syntactic choices. For example, the word *slovenly* has stronger emotional overtones than *messy*, and the agentless sentence 'Mistakes were made' is more neutral than 'The government made mistakes'.

Besides these direct linguistic expressions of emotion, both speech and writing offer additional opportunities for conveying emotion. In face-to-face communication,[2] emotion is commonly transmitted through such paralinguistic tools as facial expression, body posture, or physical proximity to our interlocutor, not to mention the volume and timbre of our voice. When we write, none of these means of expression is available. To compensate, we sometimes rely upon punctuation marks for emphasis (e.g. !!! or ???), use capital letters, underline words, put passages into bold or italic fonts, employ out-sized point type, or repeat words that, in speech, would have carried distinctive articulation (e.g., spoken: 'I'm REALLY sorry to inconvenience you' versus written: 'I'm really, really, really sorry to

---

1    I am grateful to Leopoldina Fortunati and Jane Vincent for their insightful comments on earlier drafts of this chapter.
2    Following convention, I use the term 'face-to-face' in this chapter. However, Fortunati (2005) argues that a more accurate term is 'body-to-body'.

inconvenience you'). Historically, additional contrivances included opting for a certain type of stationery, sprinkling perfume or enclosing dried flowers, or choosing a particular stamp for the envelope.

Use of devices such as repeated punctuation or forty-point type when we write suggests that our writing is attempting to mirror the emotions we might express paralinguistically in face-to-face conversation. Such devices are familiar in handwritten letters and in word-processed documents, especially in casual compositions exchanged between friends. As computer-based communication went online (initially through email, and then through listservs, newsgroups, and chat),[3] these same devices have appeared – especially in informal communication. While they are technically forms of writing, most varieties of online communication have often been thought of as forms of speech, with creative punctuation and typography substituting for paralinguistic cues (such as volume, proxemics, and facial expression) for expressing emotion.

But playing with punctuation and typography are not the only tools available for expressing emotion when writing online. Beginning in 1982, a new form of expression markers, known as emoticons, began to emerge. These markers were explicitly created with the goal of clarifying the emotion that an online writer was intending to convey in his or her message. Implicit in the creation of the early emoticons was the assumption that online communication was essentially a written version of casual speech, rather than more traditional written language (akin to a brief essay or a letter). However, since the language was physically written, the paralinguistic cues of face-to-face speech were missing, and it was presumed that the risk of misinterpretation was therefore high. (Note that in the early days of computing, users could not easily manipulate the appearance of type to add emotional nuance to their messages. Tools for doing underlining, bold and italics, or for manipulating font style or size only appeared with the development of word processing programmes, and are even more

---

3    While users face similar challenges in expressing emotions with mobile devices, this
     chapter focuses on computer-based communication, for which emoticons were originally
     invented.

recent in online communication programmes such as email and instant messaging.)

This chapter explores the history of early attempts to create written expressions of emotion in computer-mediated communication. While the original goal of such innovation centred on clarifying linguistic intent (that is, devising substitutes for traditional paralinguistic markers of emotion), we will see that the story of emoticons is also the story of assumptions people have made about whether online communication is a form of speech or a form of writing.

## Enter the Smiley

On 19 September 1982, Scott E. Fahlman sent the following message to an online bulletin board at Carnegie Mellon University:[4]

> 19-Sep-82 11:44   Scott E Fahlman         :-)
> From: Scott E Fahlman <Fahlman at Cmu-20c>
>
> I propose that the following character sequence for joke markers:
>
> :-)
>
> Read it sideways.  Actually, it is probably more economical to mark things that are NOT jokes, given current trends.  For this, use
>
> :-(

To understand why Fahlman made these suggestions, some context is useful. Networked computing got its start at the end of the 1960s, when the US Department of Defense launched its Advanced Research Projects Agency Network, known as ARPANET. During the 1970s and into the early 1980s,

---

4   <http://www.cs.cmu.edu/~sef/Orig-Smiley.htm> Accessed 24 May 2008.

ARPANET grew, but so did so-called bulletin board systems (BBSs), which enabled individual users to communicate with groups of people at a time – sometimes through ARPANET connections but also by telephone dial-up at home. In 1982, the Internet (which evolved from ARPANET) did not yet exist; that would happen in 1983. Email had been invented in 1971 (as was early computer conferencing), but the World Wide Web would not appear until the early 1990s.[5]

When Fahlman invented the smiley, the cadre of people using networked computing for interpersonal communication was comparatively small, probably numbering in the thousands. One history of the Internet indicates that in 1981, there were 213 host computer servers connecting to ARPANET. A decade later, that figure had grown to a million.[6] By 2007, over 50 million active server sites existed worldwide.[7] The International Telecommunication Union reports that as of 2006, there were 1.13 billion users of the Internet.[8] Early users communicating via networked computers had their share of challenges when it came to inputting text. The sophisticated word processing programs that today we take for granted were still in their infancy. The earliest modern program, WordStar, which was released in the late 1970s, lacked such niceties as spell-check and even the ability to type bold or italics directly onto the screen. (These were represented with embedded codes, and only appeared when a document was printed out.) WordPerfect made its debut on mini-computers in 1982, with a version for personal computers appearing a year later. Microsoft Word was also released in 1983.[9]

In the meantime, text for online communication was commonly inputted using generic text editors designed for doing programming in such

5    For a review of historical developments in networked computing and in online communication systems, see Abbate, 1999 and Baron, 2003.

6    <http://www.originami.com/sp/milestones.htm> Accessed 24 May 2008.

7    <http://news.netcraft.com/archives/web_server_survey.html> Accessed 24 May 2008.

8    <http://www.itu.int/ITU-D/ict/statistics/at_glance/KeyTelecom99.html> Accessed 24 May 2008.

9    See Bergin, 2006, for a history of early word processing.

computer languages as FORTRAN and BASIC. These editors were cumbersome and non-intuitive. Moreover, computer terminals of the time that were connected to mini-computers or mainframes were often difficult to read and permitted only a relatively small number of characters on a screen. For example, the ADM3A terminal (common in the early 1980s) could display 80 characters per line, but only 12 lines per screen.[10] When personal computers entered the commercial market, the number of characters that fit onto a screen was also initially rather small. Radio Shack's TRS 80-Model I, which appeared in 1977, could hold 16 lines of 64 characters each. The Apple II (available in the same year) permitted 24 lines, though each line held a maximum of 40 characters.[11]

We should also remember that back in the early 1980s, the United States had not yet become a nation of typists.[12] Outside of secretaries, journalists, and some authors, most people either did not know how to type or typed only with difficulty. Autodidacts in the academic community (professors and students alike) often relied on the hunt-and-peck method, using whatever fingers worked for them. This was the computing milieu in universities when the smiley (and frowny) made their debut.

Like most inventions, Fahlman's original emoticons were created to solve a concrete problem, in this case, at Carnegie Mellon.[13] A number of colleagues had been engaging, over a period of days, in casual banter on an online bulletin board, posing hypothetical questions about what the physical consequences might be if someone cut a building's elevator cable: What

10    <http://www.tentacle.franken.de/adm3a/> Accessed 24 May 2008.
11    By 1982, The IBM PC, which was now designed to handle serious word processing, offered 25 lines of 80 characters each. Information regarding textual input on mainframe, mini and personal computers of the time was graciously provided by Tim Bergin, personal communication, 21 and 26 November 2007.
12    In a more extensive discussion of how emotion is expressed through writing, it would be interesting to compare differences between handwritten documents (especially personal letters) and their typewritten or word-processed counterparts.
13    Fahlman has discussed the incident in a section of his home page at Carnegie Mellon <http://www.cs.cmu.edu/~sef/> as well as on an interview on 26 September 2007 with the Canadian Broadcasting Corporation ('The Current' – <http://www.cbc.ca/thecurrent/2007/200709/20070926.html> Accessed 24 May 2008).

would happen if there were a helium balloon in the elevator? A pigeon?
A lit candle? A drop of mercury? Continuing in this vein, the following
message was posted, with the subject line 'WARNING!':

> Because of a recent physics experiment, the leftmost elevator has been contaminated
> with mercury. There is also some slight fire damage. Decontamination should be
> complete by 08:00 Friday.

Of course, there was no mercury spill, and no fire damage. But apparently someone reading the notice had taken the posting seriously. The next
day, discussion on the bulletin board turned to the fact that in reading
an online discussion, it was easy to mistake humour for reality. Fahlman
explains that

> Given the nature of the [computing] community [at Carnegie Mellon], a good many
> of the posts were humorous (or attempted humor). The problem was that if someone
> made a sarcastic remark, a few readers would fail to get the joke, and each of them
> would post a lengthy diatribe in response. That would stir up more people with more
> responses, and soon the original thread of the discussion was buried.[14]

Contributors to the Carnegie Mellon bulletin board tried more humour
('Just when you thought it was safe to go back into the elevator ...') but also
proposed adding symbols for indicating when a posting was to be taken
as a joke. Ideas included an asterisk (*), a percent sign (%), an ampersand
(&), or a pound sign (#). Fahlman suggested using what we now know as
the smiley, i.e. :-) 'for joke markers' and what we know now as the frowny,
i.e. :-( 'to mark things that are NOT jokes'.

14    <http://www.cs.cmu.edu/~sef/sefSmiley.htm> Accessed 24 May 2008.

# Evolution of Emoticons in Computer-Mediated Communication

Within short order, the idea of inserting concatenations of punctuation marks as non-verbal comments spread across the United States: first through Carnegie Mellon, then to other universities and research sites, and eventually to the general computer-using public. In the process, new symbols were added, and the meanings of the original smiley and frowny evolved as well.

## From 'The Elevator Has Been Contaminated' to 'Have a Nice Day'

The most obvious of the early changes was a redefinition of Fahlman's proposed use of the frowny. Instead of indicating 'take this [previous comment] seriously', the mark soon came to denote 'displeasure, frustration, or anger'.[15] Shift in meaning is, of course, common in everyday language. Sometimes words change their senses entirely, as when *silly* went from signifying 'blessed' to its current meaning of 'foolish'. In other instances, language users have lost track of the meanings earlier associated with words or phrases. A *pen knife* was, originally, a pocket knife used for sharpening quills for writing. The phrase 'How do you do?' is still common when being formally introduced to someone we do not know, but in using these words, we are hardly inquiring how that person is actually doing. In fact, when we say 'How are you doing?' to an acquaintance we pass on the street (as we each hurry along in opposite directions), we frequently hope for a quick reply of 'Fine' rather than an honest answer to the question, which might slow us both down.

Meaning change also is evidenced in terminology commonly seen as typical of computer-mediated communication. The acronym 'LOL' (originally standing for 'laughing out loud') is fairly common in the instant

---

15    <http://www.cs.cmu.edu/~sef/sefSmiley.htm> Accessed 24 May 2008.

messaging conversations of American undergraduate students. However, its use often has nothing to do with humour. Instead, it functions as a phatic marker, meaning the equivalent of 'OK', 'Cool', or 'Yeah', as seen in the following example (Baron, 2004):

> MARK: i've got this thing that logs all convos [=conversations] now
> JIM: really?
> JIM: why's that
> MARK: i have ever [=every] conversation i've had with anybody since the 16th
> MARK: i got a mod [=module] for aim [=America Online Instant Messenger], and it just does it
> MARK: i'm not sure why
> JIM: lol
> JIM: cool

Over the years, the smiley has likewise undergone considerable evolution. In fact, the story predates Fahlman's 1982 proposal (Crampton, 2006; Nordin, 2006). In 1963, a graphic artist named Harvey R. Ball created a yellow button, bearing a now-familiar smiling face, for the State Mutual Life Assurance Company of America. The organization was looking to relieve tension among its employees in the midst of a merger with another company. The design (which Ball never trademarked) soon gained popularity through the efforts of Philadelphia card-shop owners Murray and Bernard Spain. The Spain brothers marketed a range of products such as coffee mugs, buttons, and bumper stickers bearing Ball's yellow smiley face, coupled with the words 'Have a happy day'. Over time, the phrase was commonly edited to 'Have a nice day'. Ball and the Spain brothers intended the smiley to denote happiness and well-being; Fahlman's online smiley face was designed to mark a preceding textual comment as being meant in jest. But soon the offline and online worlds merged. Software programs such as Microsoft Word and America Online's Instant Messenger began to automatically convert smileys (and frownys) that were created with individual keystrokes into graphic icons, i.e. ☺ and ☹. In the case of the smiley, it is hardly surprising that the general public began associating Fahlman's original symbol not just with humour but also with pleasant thoughts, e.g. 'Vacation starts tomorrow ☺ [= *I* am happy about having a vacation]' or 'Hope you feel better ☺ [= I hope *you* become happier / feel better]'.

In the process, the smiley (to take just one example) came to have multiple possible meanings. One of the main arguments for using emoticons in computer-mediated communication is that they ostensibly clarify meaning, given the absence of facial and gestural cues. However, whenever a word (or symbol) has multiple meanings, the possibility for misinterpretation arises. Consequently, it is hardly a foregone conclusion that emoticons promote – rather than hinder – unambiguous communication.

One of the particular challenges of online communication is that there is no generally-accepted rule book containing linguistic conventions that everyday users should follow (Baron, 2002). Instead, as people have gone online, they have tended to imbibe popular lore about how the medium works: Be carefree about spelling, punctuation, and grammar; use numerous abbreviations and acronyms; expect flaming (that is, rude language); and add emoticons to clarify meaning. Empirically, this characterization may have suited online communication of the 1980s and perhaps the early 1990s. However, as the number of online users exploded over the past decade, many of these practices are no longer followed, at least not by a substantial number of people. Yet myths die hard, and even those not using these early features of online communication 'know' that they characterize the medium.

As the meanings (and uses) of emoticons have evolved, so have the online platforms in which people communicate with one another. In the early 1980s the most common form of online communication was one-to-many. That is, individual users would post comments to an online bulletin board, an online computer conference, or later a newsgroup, listserv, or chat room. These comments could be read by a broad audience, many of whom the individual who was doing the posting might not know. Although one-to-one communication between two specific people was possible through email and instant messaging, availability of these tools was largely restricted to research institutions (or to businesses) until the 1990s. Not until the prices of computers (and computer networking) fell in the second half of the 1990s did one-to-one communication between people who knew each other become a dominant form of online exchange.[16] When communicat-

---

16 More recently, blogs and social networking sites have once again popularized one-to-many online communication.

ing with acquaintances (whether face-to-face, on the telephone, through letters, or online), the chances of our interlocutor misunderstanding us are smaller than when exchanging information with strangers.

## Research on Emoticons

Sustained research on language used in online communication dates back to the early 1990s (for example, Lea and Spears, 1992; Markus, 1994; Herring, 1996). A number of studies have explored the assumption that emoticons are either necessary – or at least valuable – tools that 'compensate' for the lack of social and affective cues online that characterize face-to-face communication.

Historically, it was argued that 'When there is no visual image of participants, it is often necessary to include verbal cues (e.g., emoticons) and verbal clarifications (e.g., IMHO ['In My Humble Opinion']) to convey to readers a sender's intended tone and meaning' (Kavanaugh et al., 2005; also see Reid, 1995; Thompson and Foulger, 1996). This assumption received initial support from research published in 1992 by Martin Lea and Russell Spears, who reported that such paralanguage 'is one means by which social information is communicated in [computer-mediated communication]'. They noted, however, that 'the meaning of paralinguistic marks is dependent on the group or individual context that is pre-established for the communication' (Lea and Spears, 1992, p. 321).

Andrea Kavanaugh and her colleagues further emphasize the importance of context in assessing the role of emoticons or acronyms (such as 'IMHO' or 'LOL') in online communication. Kavanaugh studied the Blacksburg Electronic Village, a project in the 1990s that 'wired' the town of Blacksburg, Virginia, and then encouraged citizens to make use of online resources. Analyses of how residents interacted online suggest that paralinguistic cues did not seem to be as necessary for clarifying meaning when communicating with friends as when messaging strangers:

> While there is still a need for such verbal cues online in networked geographic communities, the fact that most people already know each other at least as acquaintances,

provides a lot of background information about personality and manner from prior face-to-face interactions ... The fact that members of groups interacting online typically already know each other in networked communities mitigates against some of the problems of social presence online (Kavanaugh et al., 2005).

Perhaps equally importantly, the users being studied were mature adults, who were presumably already practiced at composing clear communication, e.g. through experience with writing letters or memoranda. If, under some circumstances, emoticons and acronyms contribute to clarifying meaning online, it is important to acknowledge that there may be additional, though often subtle, paralinguistic cues as well. These may include choosing to delay one's response to a message (thereby sending a negative linguistic cue) (Walther and Tidwell, 1995) or letting message length and content signal the degree of intimacy (Soukup, 2000).

Work by Joe Walther and Kyle D'Addario in 2001 began calling into question the extent to which users of online communication rely upon emoticons to interpret messages they receive. The authors report that in one-to-one communication, users depend more heavily upon the verbal content than upon emoticons to interpret the meaning of messages. Other studies of online behaviour have noted that both use of emoticons – and reliance upon them to interpret textual meaning – are subject to the same kinds of parameters as we see in face-to-face communication. Tom Postmes and his colleagues conclude that use of paralanguage varies from one online group to another, and that the same individuals will vary their behaviour, depending upon the group in which they are participating at the time (Postmes et al., 2000). Kristin Byron and David Baldridge (2007) have found that personality is another relevant factor determining the extent to which nonverbal cues are incorporated into online messaging.

Finally, a growing number of studies are exploring online communication in cross-cultural context (Danet and Herring, 2007). Some of the most interesting work with emoticons has compared American symbols with the use of *kaomoji* in Japan. While American emoticons are read sideways and emphasize the mouth, Japanese *kaomoji* are read horizontally and focus on the eyes. For example, the closest equivalent to the American smiley, that is :-), is the *kaomoji* ^--^. Studies comparing emoticons and *kaomoji* indicate that cultural differences between Japan and the US are reflected in the ways Japanese interpret these two forms of nonverbal online expression (Pollack, 1996; Yuki et al., 2007).

## Are Emoticons Necessary? Contemporary Claims

Despite growing research to the contrary, public perception that online language is especially prone to ambiguity (and that emoticons help clarify meaning) remains strong. Two assumptions seem to underlie this stance: one involving message length and the second, the status of online communication as a substitute for speech rather than writing.

### The 'Length' Argument

The first rationale is that because messages are short, it is difficult to convey meaning clearly. In an interview marking the twenty-fifth anniversary of the smiley, Fahlman articulated this position:

> Using emoticons is kind of a quick and lazy thing. Certainly it's better if you want to sit down and write a long essay about what you think. [But] there's a time and place in the world for quick [messages] ... There are alternative ways [than emoticons] that are probably better ways if you want to put the work in, if you've got time to write a longer message.[17]

In the early 1980s, online messages were commonly quite brief, driven in part by challenges in screen-size, online editing capabilities, and typing skills of the time. But is the 'length' argument valid today?

---

17    <http://www.cbc.ca/thecurrent/2007/200709/20070926.html> Accessed 24 May 2008.

## *The 'Speech' Argument*

Since the inception of platforms such as email, bulletin boards, computer conferencing, and instant messaging, researchers and users alike have
debated the linguistic nature of online messaging: Is it a form of written
or of spoken language?[18] Whether online communication is perceived as
writing or speech has important ramifications for the way users formulate
their messages, along with how people view the medium's shortcomings.
If we think of online postings as written language, then the older tradi-
tions of letter-writing, essay-writing, and such shape our expectations of
how to formulate electronic messages. However, if the mental paradigm is
that of speech, users understandably notice the lack of visual and auditory
cues that are available in face-to-face encounters, and may feel the need
to compensate. The popular press (along with many researchers and end-
users) has generally assumed that online communication is a substitute for
spoken language, and that emoticons are an important tool for filling in
the paralinguistic gaps. Is this assumption justified?

## *Responding to the 'Length' Argument*

Is there a relationship between length of a written message and clarity?
Writing, by its very nature, affords people the opportunity to reflect upon
their texts before 'publishing' them – in the literal sense of 'making public'.
Such reflection contrasts (at least paradigmatically) with informal spoken
discourse, in which words are often uttered without much forethought,
but semantic repairs can be made as the conversation proceeds.[19] As with
speech, written discourse may be concise or verbose. In the case of (written) online messages, the amount of text that writers produce has historically been comparatively brief. Our question is whether such messages are
inherently ambiguous (given their brevity) and whether emoticons help

---

18   See Baron, 1998, for a summary of the discussion as of the late 1990s.
19   See Baron, 1981, for an extended comparison of speech and writing.

disambiguate meaning. From the inception of emoticons in the early 1980s, the typical answer has been 'yes' on both counts.

People who have spent appreciable time working with high school or university students on their writing will likely find the length argument puzzling. Many of us devote considerable energies attempting to explain that lucid writing is the result of thought – and often, revision – not of piling on more words. In fact, concision has often been seen as a hallmark of good writing. We think of memorable lines from Shakespeare but also of haiku poetry, which has been the province not just of literary masters but of the larger populace. In the United States, a growing emphasis was put on simple, unadorned writing, following the American Civil War (Wilson, 1966). The apotheosis of this movement is often taken to be the writing of Ernest Hemingway, which was partially shaped by limitations on the length of telegraph messages. Hemingway spent many years as a journalist, sending newspaper dispatches over the wire, and it has been argued that telegraphic constraints on word-count influenced his later fiction (Blondheim, 1994). However, concision has been the watchword in American prose ever since.

Clear writing takes time – and, more often than not, effort. For many users of online communication, the most important characteristic of the medium is that it is quick. Rapidly composed missives often end up being short with respect to number of characters, but they also tend not to be thought through and, in many instances, are not read through before being sent. In this sense, they bear one of the fundamental traits distinguishing between our paradigmatic notions of spoken and written language: While speech is generally extemporaneous, writing typically demands more reflection before we release it for others to read. That is, the 'length' argument for using emoticons in computer-mediated communication may actually be reducible to the argument that online writing, like extemporaneous speech, often lacks deliberative forethought or post-editing.

Before leaving the 'length' argument, it is important to keep in mind that the nature of online communication has undergone radical transformation since the appearance of the smiley a quarter-century ago. The number of people who regularly use computers for online communication has grown exponentially, from a couple of thousand to hundreds of millions. (Recall that over a billion people use the Internet.)

The medium's functions have also expanded vastly, now including formal memoranda, contract negotiations, and job applications – many of which demand a careful eye. The writing skills (and standards) of this expanded usership run the gamut from very basic to highly articulate. But education level or literary abilities are not the only determinants of online style. There is considerable variation among individuals as to the amount of care they invest in composing email. Some accomplished writers view online exchange with friends as something casual, dashed off quickly, not to be proofread – unlike their elegant prose that appears in traditional print. Other people (even those without literary accomplishments or even professional aspirations) are noted for composing well-crafted emails. In short, online communication is not inherently sloppy or edited, ambiguous or clear, benefiting from use of emotions or not. Rather, individual users chose the amount of effort they wish to invest in the messages they send. Length and clarity are the result of writers' skills and choices, not of the medium itself. Some very short messages are extremely clear. Some meandering messages remain hopelessly muddled and ambiguous.

## Responding to the 'Speech' Argument

If message length by itself is not a necessary factor driving use of emoticons in online communication, what about linguistic modality? We have just noted that speech is often more hurriedly composed than traditional writing. Accordingly, we might assume that speech is more open to misinterpretation than traditional writing.[20] To complete the argument: If online language is like speech, then emoticons might be needed to rescue

---

20  Admittedly, speech (at least face-to-face) carries additional paralinguistic information. Furthermore, dialogue affords the opportunity for semantic clarification. However, if we simply compare words produced in spoken versus written expression, the latter affords more opportunity for editing before being rendered public.

us from our lack of verbal clarity, due to being in a rush (at least in comparison with writing).

A problem with this explanation is that semantic confusion in human
communication is hardly restricted to language constructed on the fly.
Miscommunications – in either writing or speech – often result not from
linguistic sloppiness but from the fact that interlocutors come to a 'conversation' (as readers or as listeners) with different levels of expressive mastery,
along with different personal histories, judgments, and presuppositions.

Consider this example from my own experience. I was sitting with a
colleague at the conclusion of a conference we had both helped organize.
As we reviewed what had gone well and what we might like to change
for next year's event, my colleague declared, 'That was a terrible business
luncheon.' I whole-heartedly agreed. However, it was only months later
that we discovered he had been complaining about the food served, while
I had found the meeting's agenda to be extremely boring. Our problem
was not being sloppy in our language. Rather, the phrase 'terrible luncheon' bore multiple meanings. Had our discussion taken place online, had
my colleague even added a smiley or frowny to the end of his message, the
ambiguity would have persisted.

A more notorious example of misunderstood intent is Orson Welles'
pre-Halloween radio broadcast of 'War of the Worlds' in 1938. The production, based on H. G. Wells' 1898 story of a Martian invasion, was scripted
as a series of news bulletins. At the beginning of the programme, listeners
were informed that what they were about to hear was fiction, with reminders
inserted both during and at the end of the show. Yet like Fahlman's tale of
the elevator at Carnegie Mellon, the story was taken by some at face value,
especially by listeners who missed the opening disclaimer – and by all the
friends, neighbours, and family they telephoned who had not been listening
in. The general public panicked. The Federal Communications Commis-
sion even launched an investigation of CBS radio, which had broadcast
the programme.[21] The moral of the story would seem to be that even three
disclaimers during an hour's broadcast were of little consequence in the face

21    See Koch, 1970, and <http://www.war-ofthe-worlds.co.uk/> Accessed 1 June 2008.

of a convincing dramatization. Just so, would a single smiley face at the end of the Carnegie Mellon elevator message necessarily have been heeded by readers who had not been following the prior discussion thread?

Most discussion involving the spoken versus written nature of online language revolves not around haste or even around linguistic ambiguity, but around the paralinguistic cues (including facial expression, bodily stance, tone of voice) that accompany face-to-face speech. The argument generally goes like this: Online language is a representation of speech. Because some of the elements that bear meaning in face-to-face communication are missing online, such media as email, instant messaging, and chat are prone to being misunderstood. Emoticons are an attempt to fill in some of the paralinguistic gaps.

The argument hinges on the assumption that users of online communication are attempting to represent speech. True, we refer to instant messaging 'conversations,' and students tell friends they will 'see them online'.[22] However, a more accurate way of assessing the veracity of the online-language-as-speech claim is to analyze the linguistic character of online language itself.

Linguistic studies have compared the prototypical features of spoken and written language.[23] Figure 1 summarizes a number of the main distinctions:

22 Baron, 2004.
23 For example, Tannen, 1982a, 1982b; Biber, 1988; Chafe and Danielewicz, 1987; Chafe and Tannen, 1987; Crystal, 1995; Baron, 2000.

|                                            | Speech                                                                             | Writing                                                                         |
| ------------------------------------------ | ---------------------------------------------------------------------------------- | ------------------------------------------------------------------------------- |
| **Structural Properties**                  |                                                                                    |                                                                                 |
| • Number of participants                   | Dialogue                                                                           | Monologue                                                                       |
| • Durability                               | Ephemera (real time)                                                               | Durable (time independent)                                                      |
| • Level of specificity                     | More vague                                                                         | More precise                                                                    |
| • Structural accoutrements                 | Prosodic and kinesic cues                                                          | Document formatting                                                             |
| **Sentence Characteristics**               |                                                                                    |                                                                                 |
| • Sentence length                          | Shorter units of expression                                                        | Longer units of expression                                                      |
| • One word sentences                       | Very common                                                                        | Very few                                                                        |
| • Sentence-initial coordinate conjunctions | Frequent                                                                           | Generally avoided                                                               |
| • Structural complexity                    | Simpler                                                                            | More complex                                                                    |
| • Verb Tense                               | Present tense                                                                      | Varied (especially past and future)                                            |
| **Vocabulary Characteristics**             |                                                                                    |                                                                                 |
| • Use of contractions                      | Common                                                                             | Less common                                                                     |
| • Abbreviations, acronyms                  | Infrequent                                                                         | Common                                                                          |
| • Scope of Vocabulary                      | More concrete<br>More colloquial<br>Narrower lexical choices<br>More slang and obscenity | More abstract<br>More literary<br>Wider lexical choices<br>Less slang and obscenity |
| • Pronouns                                 | Many 1st and 2nd person                                                            | Fewer 1st or 2nd person (except in letters)                                     |
| • Deictics (e.g. *here, now*)              | Use (since have situational context)                                              | Avoid (since have no situational context)                                      |

Figure 1: Major Distinctions between Spoken and Written Language.

Our question now is whether online communication more closely resembles speech or writing, as defined by the kinds of parameters noted in Figure 1.

In the late 1990s, I surveyed the relevant literature on email, bulletin boards, and computer conferencing, concluding that computer-mediated language was essentially a mixed modality.[24] It resembled speech in that it was largely unedited, it contained many first- and second-person pronouns, it commonly used present tense and contractions, it was generally informal, and online language could be rude or even obscene. At the same time, messages composed and transmitted online looked like writing in that the medium was durable, and participants commonly used a wide range of vocabulary choices and complex syntax (Baron, 1998).

A few years later, in his book *Language and the Internet*, David Crystal investigated a variety of types of computer-mediated language, including the web, email, chat, and virtual worlds such as MUDs and MOOs. He compared these platforms against his own analysis of spoken versus written language. Coining the term 'Netspeak' to refer to the whole of language used online, Crystal concluded that 'Netspeak has far more properties linking it to writing than to speech [...] Netspeak is better seen as written language which has been pulled some way in the direction of speech than as spoken language which has been written down' (Crystal, 2001, p. 47).

In 2003, I undertook a study of instant messaging (IM) conversations constructed by American college students. My goals were to gather empirical data on the spoken versus written nature of the medium, as well as to see whether gender was a relevant variable. Here are some highlights of my findings:[25]

---

24 My own study benefited greatly from work by Milena Collot and Nancy Belmore (1996), and by Simeon Yates (1996), whose detailed comparisons of computer-mediated conversations with large-scale corpora of spoken and written language (including those collected by Douglas Biber, the London-Lund speech corpus, and the Lancaster-Oslo/Bergen written corpus) helped shape my thinking.
25 Details of the study appear in Baron, 2008, Chapter 4.

- spelling was remarkably good
- punctuation (including capitalization, question marks, and periods) was generally consistent enough to make meaning clear
- considering how frequently contractions (e.g., *can't* instead of *cannot*) appear in informal speech, there were fewer contractions in the IM conversations than anticipated
- some of the vocabulary and syntax used was quite sophisticated

All of these features are more suggestive of a written rather than a spoken model for the discourse. When it came to emoticons, they turned out to be rather scarce. Out of 11,718 words in the corpus, only 49 emoticons appeared. Of these, 31 were smileys and 4 were frownys. Apparently, students engaged in these IM conversations did not feel the need for frequent emoticons to clarify their meaning.

My study included a number of other linguistic components, including frequency of abbreviations and acronyms. Both abbreviations (e.g., 'k' for 'OK') and acronyms (e.g., 'LOL' for 'laughing out loud') are generally assumed to be common features of online communication. In fact, though, they were about as sparse as emoticons. Out of 11,718 words, only 31 were abbreviations, of which 16 were 'k' for 'OK'. There were 90 cases of acronyms, which included 76 instances of 'LOL' – many of which served as phatic fillers (substitutes for 'Yeah' or 'Cool') rather than actual indicators of humour.[26]

I also investigated how the students 'chunked' their messages into multiple, seriatim transmission, such as

> transmission 1: that must feel nice
> transmission 2: to be in love
> transmission 3: in the spring
> transmission 4: with birds chirping
> transmission 5: and frogs leaping

---

26  Tagliamonte and Denis, 2008, report similar meaning variations with 'LOL'.

The analysis of how IM conversations are typically 'chunked' is rather complex, and I will only note the essential findings.[27] When I examined the grammatical point in a sentence at which IM messages were broken into multiple transmissions, males were more likely to put breaks at the same places typical of pauses in spoken language, while female breaks more closely resembled the places in which punctuation is used in written language. Interestingly, when I re-examined the contraction data by gender, I found that male students used contractions 77 per cent of the time the language allowed (e.g. using *can't*), while females only used contractions 57 per cent of the time (e.g. *cannot*). On the whole, then, male instant messaging conversations tended to have more in common with face-to-face speech, while female IM discourse had more in common with conventional writing. That said, neither group made much use of emoticons.

What can be concluded about the status of online communication more closely resembling spoken or written language? The question does not have a simple answer. Gender seems to play a role, as does age. Older users (including young adults) generally use fewer of the features we associate with electronic communication – such as emoticons – than do younger people (especially middle-school aged children).

But there is another important factor as well: maturation of users within the medium. During conversations I had with college students regarding instant messaging practices, I repeatedly heard how their online habits had evolved since they initially began using IM as teenagers. Earlier they might have intentionally included emoticons, abbreviations, and acronyms, in part because it felt socially appropriate to adopt a special lingo online. However, as they increasingly found themselves using computers to write papers for school, for doing online searches – in short, for getting real work done – their IM style evolved. It was now simply easier and more natural to write reasonably coherent, reasonably spelled, reasonably punctuated, largely acronym-free prose, which (particularly for the females) started to resemble the writing they might hand in for a school assignment.

27    Again, see Baron, 2008, Chapter 4 for the fuller analysis.

## Domestication of Online Communication

When new technologies are introduced, there is typically a transitional period until the general populace figures out how the technology works and comes to feel comfortable with it. Consider the case of landline telephones, which appeared in 1876. As late as 1894, a newspaper editor in Philadelphia was warning his readers not to speak on the telephone with people who had communicable diseases, for the illness might be conveyed across the telephone wires.[28] Today we find such concern laughable.

Roger Silverstone and Leslie Haddon (1996) have used the term 'domestication' to describe the process whereby a new device (such as an automobile, a vacuum cleaner, or a computer) becomes a normal part of daily living. Today, for instance, we are witnessing domestication of mobile phones. When mobiles are initially introduced into a community, users often speak loudly, unsure that their voices will carry on such a small, portable device. (A century ago, George Bernard Shaw commented on how people shouted unnecessarily into landline phones.) In Sweden, some contemporary mobile phone users comment that they speak more softly on mobile phones than when conversing face-to-face. And even Americans, who are noted for their loud voices in public, seem to be lowering their volume somewhat when talking on mobile phones, now that the technology is becoming increasingly familiar.

Online communication has been evolving as well. When email got its serious start in universities and research institutions in the 1980s, people commented on how informal and unedited email messages tended to be. Today, a lot of casual electronic communication has moved to instant messaging (on computers) or text messaging on mobile phones. University students in both the United States and Sweden have reported to me that they view email as a formal medium, which commonly merits both preplanning and proofreading.

---

28   'Diseased Germs Transmitted through Telegraph Circuits,' Electrical World, 22 June 1894, p. 833; cited in Marvin, 1988, p. 81.

The value of emoticons in computer-mediated communication has also shifted over time. When the medium was new, the community of users tended to be male university researchers whose forte was computer-related issues, not literary composition or typing skills. ARPANET was renowned both for its casual humour and for flaming, some of which likely resulted from initially innocent misunderstandings. At least in the early days, the bulk of communication was done in one-to-many forums (such as bulletin boards), where a given user might not know all the people reading the posts. Moreover, it was common for threads of conversation to spread over hours or even days. Unless you had read the entire thread, you might understandably not follow the meaning of a particular post.

The world of online communication has changed unrecognizably since 1982. Email, instant messaging, listservs, blogs, social networking sites, and text messaging on mobile phones all attract a plethora of users of both genders and spanning the range of age, education, typing skills, personality, and attitudes towards electronic communication. There is still a lot of humour online, but as often as not, it is directed to specific interlocutors who are likely to share the context necessary to understand the sender's intent.

What is more, for over a decade, word processing has been a 'killer application' for those using personal computers. The United States has indeed become a nation of typists. Cranking out long email messages is painlessly simple, given improvements in both computer screens and word-processing programs, not to mention our person-years of experience at the keyboard. As my college students can testify, it often is easier to eschew lexical shortenings (such as *U* or *can't*) in favour of *you* and *cannot*, rather than making the effort to shift back and forth between a casual online messaging style and the more standard writing typically done in Microsoft Word. With the possible exception of a whimsical smiley or frowny face, most of my students do not insert emoticons into their formal writing. Why, then, should emoticons be needed – at least for the purpose of clarifying meaning – in an email or instant message?

## Closing Remarks

Technologies evolve, and so do their users. In 1982, Scott Fahlman's proposal to use a smiley face to flag jokes on an early bulletin board system populated by a cluster of ARPANET users made sense. A large proportion of the casual postings were essentially backchannel conversation, not intended to be particularly long, carefully constructed, or serious. As the online community grew, so too did their reasons for going online and the platforms enabling them to do so. Yet as novices, people often heard via the personal grapevine – or through the press – that online missives were essentially versions of speech, that misunderstanding was common due to lack of paralinguistic cues, and that by using emoticons one could help compensate for the medium's shortcomings.

Those issuing the advice generally neglected to reflect upon the fact that ambiguity or misunderstanding can occur in any form of communication, written or spoken, and no matter how carefully crafted. What is more, few of the early, experienced users seem to have considered that the lack of paralinguistic cues in traditional written language such as letters has not caused many authors to feel the need to insert substitutes for facial expressions or bodily stances. Rather, part of the responsibility of a writer (of a novel or essay, but also of a letter to a friend) has been to find ways of encoding all of one's relevant meaning through prose.

Along with the evolution of electronic media and the people using them, the language we employ online has changed as well. Smileys now have a range of meanings (as do frownys and other emoticons). Much as the word *bad* means 'cool' in the African-American community and the word *wicked* in the phrase 'wicked good' means 'very' (not 'evil') in New England, the meanings of emoticons need to be constructed from the particular context in which they appear.[29] That is, emoticons are no more univocal than are words in ordinary language, and therefore cannot be assumed to unambiguously clarify user intention or emotion.

---

29   On the importance of situational context in interpreting lexical meaning, see Malinowski, 1923.

Now that a host of online messaging systems (such as Skype or MNS Live Messenger) offer text, voice, and video options, users can choose how much information they wish to transmit and receive. Typically, decisions are based on convenience (e.g. selecting voice rather than video because I do not want you to see I am doing other things on the computer while we are conversing) or degree of closeness to the interlocutor (e.g. students being willing to exchange textual messages with faculty members, but not talk with them, much less use video). Fear of missing out on paralinguistic cues does not seem to figure into the choice of modality.

Linguistic communication has always proved something of a paradox: Users speak (or write) to one another with the expectation of being understood, though we know that miscommunication is common. Depending upon our circumstances, we may attempt to clarify our meaning through paralanguage, rephrasing, re-editing, or insertion of graphic symbols. Scott Fahlman's introduction of smiley and frowny faces into early online communication was one such attempt, based upon assumptions about the linguistic nature of online language (i.e. as a representation of informal speech).

We have seen in this chapter that as computer-mediated communication (CMC) has evolved, the status of emoticons in clarifying meaning has become increasingly questionable. This is not to say that online communication has become increasingly unambiguous. Rather, as a form of linguistic expression, CMC is subject to most of the same challenges that spoken and written language has encountered for millennia: conveying to another person precisely what is on our mind, in all its conceptual and emotional nuance.

# References

Abbate, J., *Inventing the Internet*, Cambridge, MA: MIT Press, 1999.
Baron, N. S., *Speech, Writing, and Sign*, Bloomington, IN: Indiana University Press, 1981.

——, 'Letters by Phone or Speech by Other Means: The Linguistics of Email', *Language and Communication* (18), 1998: 133–70.

——, *Alphabet to Email: How Written Language Evolved and Where It's Heading*, London: Routledge, 2000.

——, 'Who Sets Email Style: Prescriptivism, Coping Strategies, and Democratizing Communication Access', *The Information Society* (18), 2002: 403–13.

——, 'Language of the Internet', in Farghali, A. (ed.), *The Stanford Handbook for Language Engineers*, CSLI Publications (Stanford Center for the Study of Language and Information), distributed by the University of Chicago Press, 2003: pp. 59–127.

——, 'See You Online: Gender Issues in College Student Use of Instant Messaging', *Journal of Language and Social Psychology* (23), 2004: 397–423.

——, *Always On: Language in an Online and Mobile World*, New York, NY: Oxford University Press, 2008.

Bergin, T. J., 'The Origins of Word Processing: Software for Personal Computers: 1976–1985', *IEEE Annals of the History of Computing*, October–December 2006: 32–47.

Biber, D., *Variation across Speech and Writing*, Cambridge: Cambridge University Press, 1988.

Blondheim, M., *News over the Wires: The Telegraph and the Flow of Public Information in America, 1844–1897*, Cambridge, MA: Harvard University Press, 1944.

Byron, K. K. and Baldridge, D. C., 'E-Mail Recipients' Impressions of Senders' Likability: The Interactive Effect of Nonverbal Cues and Recipients' Personality', *Journal of Business Communication* 44 (2), 2007: 137–60.

Chafe, W. and Danielewicz, J., 'Properties of Spoken and Written Language' in Horowitz, R. and Samuels, S. J. (eds), *Comprehending Oral and Written Language*, San Diego, CA: Academic Press, 1987: pp. 83–113.

Chafe, W. and Tannen, D., 'The Relation between Written and Spoken Language', *Annual Review of Anthropology* (16), 1987: 383–407.

Collot, M. and Belmore, N., 'Electronic Language: A New Variety of English', in Herring, S. (ed.), *Computer-Mediated Communication: Linguistic, Social, and Cross-Cultural Perspectives*, Amsterdam: John Benjamins, 1996: pp. 13–28.

Crampton, T., 'Smiley Face is Serious to Business', *New York Times*, 2006. <http://www.nytimes.com/2006/07/05/business/worldbusiness/05smiley.html?_r=1&oref=slogin&pagewanted=print> Accessed 5 July 2008.

Crystal, D., *The Cambridge Encyclopedia of the English Language*, Cambridge: Cambridge University Press, 1995.

——, *Language and the Internet*, Cambridge: Cambridge University Press, 2001.

Danet, B. and Herring, S. (eds), *The Multilingual Internet*, New York, NY: Oxford University Press, 2007.

Fortunati, L., 'Is Body-to-Body Communication Still the Prototype?', *The Information Society* 21 (1), 2005: 1–9.

Herring, S. (ed.), *Computer-Mediated Communication: Linguistic, Social, and Cross-Cultural Perspectives*, Amsterdam: John Benjamins, 1996.

Kavanaugh, A., Carroll, J. M., Rosson, M. B., Zin, T. T. and Reese, D. D., 'Community Networks: Where Offline Communities Meet Online', *Journal of Computer-Mediated Communication 10* (4), 2005. Available at <http://jcmc.indiana.edu/vol10/issue4/kavanaugh.html>.

Koch, H., *The Panic Broadcast: Portrait of an Event*, Boston: Little Brown, 1970.

Lea, M. and Spears, R., 'Paralanguage and Social Perception in Computer-Mediated Communication', *Journal of Organizational Computing* (2), 1992: 321–41.

Malinowski, B., 'The Problem of Meaning in Primitive Languages', Supplement 1 to Ogden, C. K. and Richards, I. A. (eds), *The Meaning of Meaning*, New York, NY: Harcourt, Brace and Company, 1923: pp. 296–336.

Markus, M. L., 'Finding a Happy Medium: Explaining the Negative Effects of Electronic Communication on Social Life at Work', *ACM Transactions on Information Systems* 12 (1), 1994: 119–49.

Marvin, C., *When Old Technologies Were New: Thinking about Electric Communication in the Late Nineteenth Century*, New York, NY: Oxford University Press, 1988.

Nordin, K., 'Smiley Face: How an In-House Campaign Became a Global Icon', *Christian Science Monitor*, 4 October 2006, <http://www.csmonitor.com/2006/1004/p15s01-algn.htm>.

Pollack, A., 'Happy in the East ^--^ or Smiling :-) in the West', *New York Times*, 12 August 1996 <http://query.nytimes.com/gst/fullpage.html?res=9905E0D8133EF931A2575BC0A960958260&n=Top/Reference/Times%20Topics/People/P/Pollack,%20Andrew>.

Postmes, T., Spears, R. and Lea, M., 'The Formation of Group Norms in Computer-Mediated Communication', *Human Communication Research* 26 (3), 2000: 341–71.

Reid, E., 'Virtual Worlds: Culture and Imagination', in Jones, S. (ed.), *Cybersociety: Computer-Mediated Communication and Community*, Thousand Oaks, CA: Sage Publications, 1995: pp. 164–87.

Silverstone, R. and Haddon, L., 'Design and Domestication of Information and Communication Technologies: Technical Change and Everyday Life', in Silverstone, R. and Mansell, R. (eds), *Communication by Design: The Politics of Information and Communication Technologies*, Oxford: Oxford University Press, 1996: pp. 44–74.

Soukup, C., 'Building a Theory of Multimedia CMC: An Analysis, Critique and Integration of Computer-Mediated Communication Theory and Research', *New Media & Society* 2 (4), 2000: 407–25.

Tagliamonte, S. and Denis, D., 'LOL for Real! Instant Messaging in Toronto Teens', *American Speech* 83 (1), 2008: 3–34.

Tannen, D., 'Oral and Written Strategies in Spoken and Written Narratives,' *Language* (58), 1982a: 1–21.

——, 'The Oral/Literate Continuum in Discourse', in Tannen, D. (ed.), *Spoken and Written Language: Exploring Orality and Literacy*, Norwood, NJ: Ablex, 1982b: pp. 1–16.

Thompson, P. A. and Foulger, D. A., 'Effects of Pictographs and Quoting on Flaming in Electronic Mail', *Computers in Human Behavior* 12 (2), 1996: 225–43.

Walther, J. B. and D'Addario, K. P., 'The Impacts of Emoticons on Message Interpretation in Computer-Mediated Communication', *Social Science Computer Review* 19 (3), 2001: 323–45.

Walther, J. B. and Tidwell, L. C., 'Nonverbal Cues in Computer-Mediated Communication, and the Effects of Chronemics on Relational Communication', *Journal of Organizational Computing* (5), 1995: 355–78.

Wilson, E., *Patriotic Gore: Studies in the Literature of the American Civil War*, New York, NY: Oxford University Press, 1996.

Yates, S., 'Oral and Written Aspects of Computer Conferencing', in Herring, S. (ed.), *Computer-Mediated Communication: Linguistic, Social, and Cross-Cultural Perspectives*, Amsterdam: John Benjamins, 1996: pp. 22–46.

Yuki, M., Maddux, W. W. and Masuda, T., 'Are the Windows to the Soul the Same in the East and West? Cultural Differences in Using the Eyes and Mouth as Cues to Recognize Emotions in Japan and the United States', *Journal of Experimental Social Psychology* 43 (2), 2007: 303–11.

# An Inconvenient Truth:
# Multimodal Emotions in Identity Construction

MARIA BORTOLUZZI

## Introduction

This chapter focuses on how emotions can be analysed and interpreted within linguistic and multimodal frameworks of analysis as used in the projects dealing with critical multimodal communication and carried out by our research unit in Udine University. The literature about the way emotions are expressed linguistically is wide-ranging because, as Fussell writes in her seminal book about this topic: '[...] the interpersonal communication of emotional states is fundamental to both everyday and clinical interaction' (Fussell, 2002a, p. 1). *The Verbal Communication of Emotion*, edited by Fussell (2002b), is particularly relevant to this chapter because it offers an interdisciplinary perspective and includes contributions about the relation between verbal and non-verbal communication, cross-cultural studies, everyday and therapeutic communication and so on. Among many other studies in linguistics, Niemeier and Dirven (1997) and Wierzbicka (1999) are cross-cultural studies of emotions expressed linguistically; Kövecses (2000) deals with metaphors conveying emotions; Pavlenko (2005) investigates the relation between emotions and multilingualism. What has not been widely investigated yet is the relation between emotions and multimodal communication, namely language, images and sounds in texts such as videos.

The main aim of this chapter is to offer an insight into a developing branch of research which has seen the linguistic field of discourse analysis expanded into the field of socio-semiotic studies to deal with multimodal meaning making. Secondly, the chapter aims at showing how the analytical

tools of this kind of research can offer fruitful insights into the investigation of emotions as expressed in mediated communication.

An introduction about the project will be followed by the outline of
the theoretical frameworks used to investigate emotions and then by the
analysis of the case-study presented here. The chapter is structured as follows: after a general outline of the research field and the literature related
to it, the case-study, the aims of the chapter and the main terminology are
presented. The data are then analysed and the findings interpreted while
the final section offers some concluding remarks.

## The General Project

The project[1] *Mediating Ideology and Negotiating Identity in Text, Image
and Sound* focuses on two key aspects of multimodal texts: their relationships with the ideology embedded in them and the identities reproduced,
reinforced, negotiated or challenged through the synergic interplay of different semiotic codes (verbal, visual, acoustic and kinetic).[2] The main tenet
at the basis of the research is that the construction of different identities
for publics and communities of practice[3] in situated actions is influenced
by overt and covert ideologies[4] conveyed by the powerful interaction of
verbal and non–verbal communication and constructed at times as non-
negotiable common sense. The project investigates how underlying ideologies become 'transparent' (i.e. not easily noticeable) when mediated through
a variety of persuasive strategies multimodally co-deployed. In these cases,

1    This project is co-ordinated by Nicoletta Vasta at Udine University.
2    Among others: van Dijk, 1998; Bauer and Gaskell, 2000; Scollon and Scollon, 2003,
     2004; Iedema, 2003; Norris, 2004; Norris and Jones, 2005; Kress and Van Leeuwen,
     2006; Lassen et al., 2006; O'Keeffe, 2006; Fairclough, 2001, 2006; Baldry and Thibault,
     2006; Vasta, 2006; Hagan, 2007.
3    Wenger, 1999; Riley, 2002; Cortese and Duszak, 2005.
4    See, among others, van Dijk, 1998, 2001, 2003, 2005; Fairclough, 2001, 2003, 2006; Weiss
     and Wodak, 2002; Martin and Wodak, 2003; Wodak, 2001, 2006.

the degree of negotiation lowers, while the critical stance the reader/user can have towards the text tends to become less effective.

The theoretical–methodological framework used to analyse the data and discuss the results will draw on systemic functional linguistics in a social semiotic perspective, including its applications to the analysis of visual grammar and multimodality in general, as well as on seminal studies in the field of Critical Discourse Analysis and Appraisal Theory.

Firstly, systemic functional grammar (Halliday, 1994; Halliday and Matthiessen, 2004; Martin, 1992; Davies and Ravelli, 1992; Hasan et al., 2005) has been adopted as the model of analysis because language:

> [...] as a means of reflecting on things and as a means of acting on things [...] is one of the semiotic systems that constitute a culture. [...] By their everyday acts of meaning, people act out the social structure, affirming their own statuses and roles, and establishing and transmitting the shared systems of value and of knowledge (Halliday, 1978, p. 2).

Secondly, the framework of analysis for multisemiotic/multimodal texts (i.e. combining verbal and non-verbal meaning-making resources which are codeployed to produce an overall textual meaning) is based on the multisemiotic studies of Kress and van Leeuwen (2001, 2002, 2006), van Leeuwen (1999, 2004), Baldry (2000, 2005), Iedema (2003) and Baldry and Thibault (2006). Iedema summarizes thus the necessity for a multi-semiotic approach:

> The trend towards the multimodal appreciation of meaning making centres around two issues: first, the de-centring of language as favoured meaning making; and second, the re-visiting and blurring of the traditional boundaries between and roles allocated to language, image, page layout, document design, and so on (Iedema, 2003, p. 33).

He also highlights: '[...] our human predisposition towards multimodal meaning making' (ibid. p. 33).

Another development stemming from systemic functional studies is Critical Discourse Analysis (CDA), the branch of linguistics and social semiotics that offers a fundamental contribution to the exploration of the two central issues of the project: ideology and identity. Van Dijk defines it:

[...] a type of discourse analytical research that primarily studies the way social power abuse, dominance, and inequality are enacted, reproduced, and resisted by text and talk in the social and political context (van Dijk, 2001, p. 352).

As van Dijk points out, CDA is not a new school, but rather it is a necessary 'critical' perspective which contextualizes, interprets and evaluates linguistic and multimodal data as never value-free but as the result of power struggles (Wodak and Meyer, 2001; Fairclough, 1992, 2001, 2003; Bora and Hausendorf, 2003; Wodak and Chilton, 2005; Wodak, 2006).

The project also refers to recent developments of Appraisal Theory which focuses on the devices used to construe (overtly or covertly) an evaluative stance in texts (White, 2005, 2006; Martin and White, 2005; see below) and the latest developments in the cognitive analysis of metaphorical expressions (Goatly, 2007; see below).

## Emotions in a Film Trailer

One of the multimodal texts belonging to the corpora we are collecting for the project is used as the case-study in this chapter. It is the trailer of the film *An Inconvenient Truth* (2006) based on the dedicated effort of Al Gore to communicate his urgent call for action against global warming.

The reasons for the choice of this trailer are firstly that it promotes a film/documentary denouncing a human problem at the global level. Ostensibly the film and the trailer belong to a wide variety of socially committed texts focused on the campaign for a more responsible attitude towards ecological issues and, in particular, to reduce global warming (the organisation promoting the film and Al Gore as its representative were awarded the Nobel Prize for Peace in autumn 2007 for their commitment to a cause of global concern). Social campaigns tend to be perceived by the general public as either 'neutral' and 'objective' or ideologically correct by default; whereas they are in fact, as all texts and discourses, the results of underlying ideological tenets. Values and identities are conveyed and construed (depending also on the communities of stakeholders directly or indirectly addressed) in ways which might be in contradiction with the ostensible aims and principles of the text. When this occurs, the ultimate purpose of a social campaign to raise the level of social awareness, critical thinking and affect social behaviour, might be partly defeated.

The second reason for choosing this film trailer is the highly emotional quality of the text as well as its inherent (and intriguing) hybridity whereby the narrative constructed by the trailer is rather different from the narrative constructed by the film it presents, since the promotional function of the text heavily influences its emotional texture. The trailer, that can be found online on the official website of the film,[5] takes some of the characteristics of the film to an extreme and becomes a highly emotional hybrid text chosen to advertise and present both the film and the DVD.

*An Inconvenient Truth* (like most film production today) well represents what Fairclough (2003, 2006) calls 'genre chains':

> [...] different genres which are regularly linked together, involving systematic transformations from genres to genres. Genre chains contribute to the possibility of actions which transcend differences in space and time, linking together social events in different social practices, different countries, and different times, facilitating the enhanced capacity for 'action at a distance' which has been taken to be a defining feature of contemporary 'globalisation', and therefore facilitating the exercise of power (Fairclough, 2003, p. 31).

In this case the film/documentary released for the big screen was edited for the DVD and contains extra features connected to the making of the film and the choices of the film-makers; the website of the film is linked to ecological activism and includes a variety of texts among which is the trailer[6]. The genre chain continues with reviews, reports and articles released in a variety of media and related to the film production and its contents. The trailer is one link of the whole, complex and in-progress chain created by the combined media phenomena 'Al Gore – *An Inconvenient Truth*'.

5    <http://www.climatecrisis.net/ > Accessed 23 May 2008.
6    <http://www.climatecrisis.net/trailer/> Accessed 23 May 2008.

The focus of this analysis relates to emotions multimodally conveyed and evoked in the trailer and, as a consequence, the image construed for the film in its 'visiting card' and advertisement (trailers are a 'promotional genre' which has the main purpose 'to arouse the viewer's curiosity and expectations about [the film]', Maier, 2006, pp. 1–2). When watching the trailer, the viewers are also construed as stakeholders belonging to different communities of practice. Multiple identities are created for the film and for the viewers by arousing emotions, involving the public and convincing them to watch the film promoted by the trailer. In trailers, the persuasive and promotional function is crucial, but their communicative results differ from other types of commercial because the advertised product is a film, that is a ready-made narrative in edited images, sounds and music which is usually the starting point and the main materials for the promotional message. As Maier writes: 'When seeing a film trailer, the viewer "tries" the product. [...] The film trailer is not an aesthetic product, but a promotional one with embedded aesthetic elements "borrowed" from another text (a film) for promotional purposes' (Maier, 2006, pp. 2–3). In other words intertextuality and interdiscursivity are at the basis of any trailer.

Intertextuality has been investigated over the years as one of the pervasive characteristics of language and it is related to what has been called by Bakhtin (1981), heteroglossic and dialogic aspects of texts. For the present study I will adopt Fairclough's definition of intertextuality (derived from Bakhtin, 1981) and the related concept of interdiscursivity, while I will extend both to visual and aural communication:

> The intertextuality of a text is the presence within it of elements of other texts (and therefore potentially other voices than the author's own) which may be related to (dialogued with, assumed, rejected, etc.) in various ways [...] (Fairclough, 2003, p. 218).

> Analysis of the interdiscursivity of a text is analysis of the particular mix of genres, of discourses, and of styles upon which it draws, and of how different genres, discourses or styles are articulated (or 'worked') together in the text (Fairclough, 2003, p. 218).

One of the most noticeable characteristic of the film *An Inconvenient Truth* is intertextuality, interdiscursivity and genre mixing (e.g. documentary of popularization of science, journalistic report, newscasting, personal narrative, and so on). The trailer takes this feature to the extreme, enhancing mainly those aspects related to heightened emotions. The main tenet of the analysis is that the device of arousing emotions in the film has the aim of personally involving the viewer and making the call-to-action more effective. The basic hypothesis of the present study is that enhancing the arousal of powerful emotions in the viewer by editing the trailer's intertextuality and interdiscursivity differently from its original source has the backlash of inscribing the film into categories which belong to fiction rather than documentary of denunciation to raise social responsibility. This may have the positive effect, ostensibly clear in the promotional function of the genre 'trailer', of attracting more viewers, but also the negative effect of importing the impression of 'unreality' or *dejà vu* typical of action and disaster movies.

## Mediated Emotions and Ideology: Working Definitions

In this section I will define and relate two main terms of the chapter: emotion and ideology where the latter tends to underlie and trigger the former in mediated communication. Ideology is a slippery and controversial notion (Manheim, 1936). Van Dijk, while recognising that ideologies are socially shaped, highlights the relevance of cognitive aspects and shared mental representations of members of a group: 'This means that ideologies allow people, as group members, to organise the multitude of social beliefs about what is the case, good or bad, right or wrong, *for them* and to act accordingly' (van Dijk, 1998, p. 8). Wodak and Fairclough have a complementary social-constructionist stand: Wodak sees ideology as 'serving the purpose of establishing and maintaining unequal power relations' (Wodak, 2001, p. 9); Fairclough remarks that '[...] ideologies are seen as one modality of power, a modality which constitutes and sustains relations of power through producing consent or at least acquiescence, power through hegemony rather than power through violence and force' (Fairclough, 2006, pp. 23–4). Fairclough insists on the dangers of commonsensical acceptance in the field of ideology: 'The ideologies embedded in discursive practices are most effective when they become naturalized and achieve the status of common sense' (Fairclough, 1992, p. 87; see also Hodge and Kress, 1993). Goatly (2007) states that non-ideological thought is impossible, but also that '[...] some ideologies of ways of understanding the world may be more useful than others. Even some ideology may simultaneously have useful and harmful effects' (Goatly, 2007, p. 1). In this respect, Goatly agrees with Balkin (1998), who remarks that ideology can be '[...] empowering, useful and adaptive on the one hand, and disempowering, distorting and maladaptive on the other' (Balkin, 1998, p. 126).

Emotion is one of the means by which ideology is conveyed in the trailer (and in the film) as instantiated by two widely used devices of mediated discourse: personalization and evaluation strategies, both closely related to the interpersonal macrofunction in systemic functional linguistics (Halliday, 1994). In recent years there has been a generalised move towards personalisation in mediated information and media discourse (Hill, 2007; Fairclough, 2003). In the trailer (and the film) information is highly personalized thanks to a variety of devices which focus around Al Gore, the main addresser/participant.

Evaluation strategies are a wide-ranging phenomenon of texts that has generated a variety of approaches (Hunston and Thompson, 2000; Maier 2006, p. 49 et passim). Appraisal Theory has been chosen for this chapter because it offers coherent and complementary insights with the other theoretical approaches and is particularly suitable for the analysis of that specific field of evaluative strategies which are related to emotions: '[...] it is a particular approach to exploring, describing and explaining the way language is used to evaluate, to adopt stances, to construct textual personas and to manage interpersonal positionings and relationships' (White, 2005). It is concerned with the linguistic means by which '[...] texts/speakers come to express, negotiate and naturalize particular inter-subjective and ultimately ideological positions' (White, 2005). It is a framework for exploring evaluative language which identifies '[...] two primary modes of evaluative positioning – the attitudinal and the dialogistic' (White, 2005).

The analysis will be integrated with the multisemiotic approach following Baldry (2000), O'Halloran (2004), Baldry and Thibault (2006) and van Leeuwen and Machin (2007).

Given the emphasis on mediated communication, I will investigate emotions using some basic findings of conceptual metaphor theory (Goatly, 2007; Kövecses, 2005; Lakoff and Johnson, 1980). One of the insights of this theory is that abstract expressions and thoughts are communicated through metaphors and metaphorical expressions are not random but fall into patterns.

As far as emotions are concerned, I will use one of the working definitions offered by psychology (Fedeli, 2006; LeDoux, 1998) whereby feeling an emotion consists in physiological, behavioural and thought alterations happening simultaneously (Fedeli, 2006, p. 14) with the following characteristics:

> [...] a) [emotions] are based on body and behaviour alterations; b) they tend to be outside the voluntary control of the individual since they are mediated by the limbic system; c) they appear before the cerebral cortex has a conscious representation of what is happening. In this sense, emotions are common to people and animals (Fedeli, 2006, p. 23; my translation).

It is interesting to notice how this and other definitions offered by neuropsychologists and scholars investigating how the brain works (Damasio, 1998, 1999) well matches most recent work done in the field of cognitive linguistics, in particular, as already mentioned, research about metaphors.

## Globalising Emotions in a Nutshell: The Film Title

I will now briefly analyse the choice of the title, which appears at the end of the trailer, and the subtitle, which never appears in the trailer: *An Inconvenient Truth. A Global Warning.* They both reveal the type of communication ostensibly at stake: the attention focuses on one *truth* which is clearly defined as generally *inconvenient* (Judgement, see the data analysis below) and, by implication, a truth that tends not to be recognized or accepted.

Thus follows the subtitles: the film is *A Global Warning* that is a direct address to the general/global public with the aim of empowering them with information and advice about the global danger we are all undergoing while calling us all for action (Judgement). The subtitle, punning with the lexical collocation 'global warming' telescopes in its 'warning' the macro ideational content: the 'inconvenient truth' in a nutshell. Right from the title we get to know what the addressers want to denounce to the general public while warning them about the global danger impending on us all. The moral stand and the emotions involved within it are already clear as are the informed position of the addressers, the pedagogical slant of the communication and the urgency of the issue.

## Voices in the Trailer

The trailer is highly heteroglossic in its use of intertextuality and interdiscursivity and it is precisely in the interaction of voices and discourses that emotions mostly emerge. The overarching genre of the film can be considered 'documentary' and the register 'popularization of science'. However, the documentary is also a 'journalistic investigation and report', a 'public lecture' (as the presence of the audience in the lecture theatre confirms) and so on. I will not linger here on the complex hybrid quality of the film, but rather focus on the way the trailer, a hybrid text by default (Dusi, 2002), conveys heightened emotions.

First of all it is possible to recognize three main and overlapping sets of addressers and their verbal and non-verbal characteristics (linguistic and paralinguistic aspects, sound and visual references associated with them). The sets of addressers and their communication have been carefully woven into the text by masterful editing. They are discussed separately because each group conveys different types of emotion in Appraisal Theory's terminology (as shown in the data analysis).

In the trailer there are three dominant ostensible addressers/narrators/voices which embed many others; these addressers are intertextual and interdiscursive clusters of voices which overlap in the editing of the trailer:

I list them below as separate, but in fact intersection of voices happens consistently.

1. The trailer's voiceover narrator (external to the action and the narrative) is represented by oral and written voices, variety of images, editing, music and it tends to blend with the intra-diegetic narrator (see below):

- Intertitles (white block capitals on black screen moving towards or away from the viewer) edited into the trailer and emphasized by silence, 'auditory' bullets or an emphatic deep male voice-over; they belong to two main types: implicitly promotional (mainly evaluative) and explicitly promotional (promotional identification and interpretation)[7]
- the editing of rapid successions of shots taken from the movie (showing the result of global warming in fast moving images);
- soundtrack (sounds and music) underlining speed, suspense and impact by means of an emphatic and fast music score reminiscent of disaster movies;
- voiceover narrator (deep, highly emphatic male voice, interdiscursively reminiscent of action or disaster movies) emphatically reading the intertitles after they appear half-way through the trailer;

2. Al Gore (narrator and main human character present in the narrative of film and trailer):

- The first off screen voice we hear in the trailer while we watch the graph of the 10 hottest years and follow the light-beam vector of the graph going upward on the background of images representing and evoking heat; at first the disembodied voice is used to focus on the scientific data presented both visually and aurally while the viewer is the position of actually 'moving up' with the vector of the increasing temperature.

---

7    I adopt here the categorisation of trailer's stages by Maier (2006, pp. 105–6).

- The scientific community: Al Gore explicitly mentions 'scientific consensus' and reports studies and data referring to climate changes, thus ostensibly embedding the authority of the scientific community in his discourse which started with a dynamic graph (the hottest years) representing scientific data;
- Al Gore the politician: 'I used to be the next President of the United States'. He defines himself politically by using temporal deixis in a deviant way (future reference 'next' with past tense) and thus triggering by implication a whole set of politically controversial events evoked in a single utterance (the year 2000 presidential elections);
- The moral voice which places the issue discussed in the film 'beyond' the political level as a global call for action; he embodies the moral commitment for a cause of urgent public concern (see point below);
- The public speaker: the trailer shows him as a successful speaker lecturing worldwide about the issue he is devoting much of his time and efforts (this is also related to his moral voice);
- The politician and media man who relates the issue of global warming to highly emotional events, their causes and results; among them hurricane devastations and crowds of refugees; also he raises the sensitive issue of terrorist threat as a comparison (see in the transcript the reference to the World Trade Centre Memorial, below).

3. The news world of the global media (woven into the heteroglossic voices of the two previous narrators):

- Overt intertextuality: televised news reels; off screen voice of journalist commenting on the images of hurricane Katrina.
- Less overt, but still clearly recognizable intertextuality and interdiscursivity in the images taken from televised news (refugees, floods), television reports (weather forecasts, reports about climate changes and pollution, etc), televised public lecture, satellite images (the earth and the hurricanes from space), and so on.

In the trailer the three sets of addressers (and their different voices, images and sounds) overlap and blend in an original and coherent hybrid which results into a heightened emotional narrative compressed into two and a half minutes. Two apparently contradictory tendencies to globalize, generalize and personalize, blend seamlessly in the final product both at the level of narrative voices (as seen so far) and at the level of construing the identity of the audience.

The message is addressed globally to the communities of practice belonging mainly to the industrialized world or to the fast developing world (China, for instance). But the message is construed in such a way as to involve emotionally the individual viewer, by placing him or her in the position of being directly addressed by these voices, as the next sections will show.

## Metaphors we have to live by

Goatly (2007) is a study of metaphor banking on recent developments in cognitive linguistics and based on large corpora of data used to compile dictionaries. He identifies the following metaphorical areas related to emotions: 'EMOTION IS FORCE, EMOTION IS MOVEMENT, EMOTION IS A CURRENT IN A LIQUID, EMOTION IS WEATHER' (Goatly, 2007, pp. 197–206). They all underline: '[...] lack of conscious control, changeability, and suddenness [...] This pejorative view of emotion is probably linked to the physiological theory that emotions are partly constituted by changes in the body which threaten its stability' (Goatly, 2007, p. 205). Damasio defines emotions as the physiological reactions to bodily states and the disturbances to them; feelings are conceptualisations or perceptions of these emotions (Damasio, 2003; see also Fedeli, 2006).

In a wider sense, 'EMOTION IS SENSE IMPRESSION (IMPACT / TOUCH > HEAT / EXPLOSION)' (Goatly, 2007, p. 224), and this is verbally substantiated in the trailer by the following expressions:

[...] standing ovations, shake you to your core; global warming; hurricane Katrina slammed into New Orleans; A film that shocked audiences; If this were to go, sea levels would go up twenty feet; Think of the impact of a hundred thousand refugees and then imagine a hundred million [transcript of trailer, see below].

The most powerful metaphors of sense impressions, however, are expressed in the fast succession of moving images edited in the trailer, in which visual metaphors of force, weather, liquid, current, impact, heat and movement blend with sounds representing and evoking impact, movement, speed and extreme weather conditions. Fast and emphatic background music reminiscent of disaster movies further increases the emotional level of communication; this alternates with shots accompanied by silence or 'auditory bullets' contributing to an atmosphere of emotional suspense.

Visual and auditory metaphors enhance the most noticeable feature of the metaphorical verbal expressions related to emotions, that is the direct involvement of the viewer as the participant who undergoes the experience ('it will shake you to your core'), while the Actor (in Halliday's grammar (1995), the participant who is responsible for setting the event in motion) tends to be the weather or the film: '[...] *hurricane* Katrina slammed into New Orleans; *a film* that shocked audiences; if *this* were to go *sea levels* would go up twenty feet' (see above).

In summary, by using a variety of semiotic codes, the trailer widely exploits the metaphor EMOTION IS WEATHER and powerfully communicates the message WEATHER IS EMOTION.

The second set of emotional metaphoric expressions is related to MORAL VALUE: *did the planet betray us or did we betray the planet? Threat; at stake.* I will deal with these metaphors by linking them to 'evaluation strategies' in the next section.

## Evaluating the Issue

As already mentioned in the previous section, the general effect of editing, visual metaphors, sound, music, voice quality is highly emotional; the co-deployed semiotic codes contribute to inscribe the target domain (the emotion created by the issue of global warming) onto the metaphorical source domain of SENSE IMPRESSION (IMPACT / TOUCH > HEAT / EXPLOSION) AND MOVEMENT, LIQUID and the most relevant of all WEATHER. Verbal communication from the three main addressers is carefully woven into the trailer through graphics, voice quality and editing to enhance the emotional involvement of the viewers but also to expand on the second set of metaphoric expressions, MORAL VALUE (mentioned at the end of the previous section).

Table 1 analyses the verbal voices (both oral and written) of the three main addressers in the trailer: the voiceover narrator (intertitles are written in block capitals as in the trailer; in small letters when spoken by the off screen narrator), the narrator/main character Al Gore and voices from the news world (all reported in quotes). The first column reports the verbal voices and some basic visual and sound information. The second and third columns are the analysis of the verbal interaction carried out using the framework of Appraisal Theory. The table reports the transcript of the voices in the trailer in quotes and the description of the visual and sound context without quotes.[8] The expressions are related to evaluative strategies and are commented on at the end of the table.

'The term "Appraisal" is used as cover-all term to encompass evaluative uses of language' (White, 2005); the verbal communication of the trailer is evaluative throughout, thus making Appraisal Theory an interesting framework of analysis. The most common device used by the voices in the trailer can be inscribed into the category 'Judgement' which is the language, '[...] which criticizes or praises, which condemns or applauds the behaviour – the actions, deeds, sayings, beliefs, motivations, etc. – of human individuals and groups' (White, 2005).

In the text, the overt Judgement expressed as a generalized statement 'Scientific consensus is that WE are causing global warming' is highlighted and sanctioned by the applause that follows and directs the viewer to identify with the audience in the lecture theatre. The exemplification of how

8    My transcript of the original trailer in <http://www.climatecrisis.net/trailer/>, accessed on 23 May 2008.

we cause global warming is achieved by editing black and white images of traffic jams and congestions in cities and roads. From then on, the reference to apparently 'neutral' weather changes are all referred to human action and therefore cannot be analysed as Appreciation (which evaluates the result itself), but rather as inscribed (overt) or invoked (not explicitly stated but rather clear in linguistic terms) Judgement of the event as specified in the third column of the table.[9]

| Verbal communication, images and sounds | Type of evaluation | positive/negative; target of evaluation |
| --- | --- | --- |
| Fast and short shots of natural disasters with accelerated movement and graphs of statistics flashed onto the screen – soundtrack evoking fast change. | | |
| Written narrator (intertitles with auditory bullets, white typeface on black screen): 'AT SUNDANCE RECEIVED THREE STANDING OVATIONS | Appreciation | + film |
| 'IT WILL SHAKE YOU TO YOUR CORE | Affect | +/– you / audience |
| 'IF YOU LOVE YOUR PLANET ... | Affect | + your planet |
| 'IF YOU LOVE YOUR CHILDREN ... | Affect | + your children |
| 'YOU HAVE TO SEE THIS FILM' | Judgement | + this film/inscribed |

9    The third column of the table also specifies whether the evaluative expression referred to tends to be positive or negative (at times it is both) in the specific context; this is represented with a plus or a minus.

| | | |
|---|---|---|
| Graph of the 10 hottest years on background of images of heat – vector of graph as light-beam going upward and followed by the video-camera.<br><br>Al Gore's offscreen voice:<br>'If you look at the ten hottest years they have ever measured, they have all occurred in the last 14 years, and the hottest of all was 2005.' | Appreciation<br>Appreciation<br>Appreciation | – years<br>– last 14 years<br>– year 2005 |
| Written narrator (intertitle with auditory bullets, white typeface on black screen):<br>'BY FAR THE MOST TERRIFYING FILM YOU WILL EVER SEE' | Appreciation | + / – film |
| Music starts – black and white images of traffic and congestion.<br>Al Gore's offscreen voice:<br>'Scientific consensus is that WE are causing global warming' [applause]. | Judgement | – global warming / inscribed |
| Al Gore's voice – images show him in a theatre in front of an audience:<br>'I am Al Gore. I used to be the next President of the United States' [laughter from audience in theatre]. | Judgement | – presidential elections / invoked |
| Pictures of places compared in the past and the present situation.<br>Al Gore's offscreen voice:<br>'This is Patagonia 75 years ago and the same glacier today.<br>'This is Mount Kilimanjaro, 30 years ago and last year. Within a decade there will be no more snows of Kilimanjaro.' | Judgement | – snows / invoked |
| Images of sufferance caused by hurricane Katrina – Al Gore's offscreen voice:<br>'This is ruling NOT a political issue so much as a moral issue.<br>'Temperature increases are taking place all over the world and that's causing stronger storms.' | Judgement<br>Judgement | – issue of global warming /inscribed<br>– stronger storms / inscribed |

| | | |
|---|---|---|
| News reels of disasters caused by hurricane Katrina.<br>Newscaster's voice 1:<br>'This is the biggest crisis in the history of this country.'<br><br>Written narrator (intertitle white block capitals on black background):<br>'IN AUGUST 2005'<br><br>Newscaster's voice 2:<br>'Early this morning hurricane Katrina slammed into New Orleans.' | Judgement<br><br><br><br><br><br><br><br>Appreciation | – meteorological crisis / inscribed<br><br><br><br><br><br><br><br>– New Orleans |
| Images of natural disasters. Written narrator (intertitles white block capitals on black background):<br>'DID THE PLANET BETRAY US ...<br><br>'OR DID WE BETRAY THE PLANET?' | Judgement<br><br>Judgement | – us /everybody / inscribed<br>– the planet / inscribed |
| Al Gore's offscreen voice – images of natural disasters:<br>'Is it possible that we should prepare against other threats besides terrorists?' | Judgement | – threat / inscribed |
| Written narrator's voice (intertitle in white block capitals on black background):<br>'FROM PARAMOUNT CLASSICS'<br><br>Narrator's offscreen voice:<br>'From Paramount Classics comes a film that has shocked audiences everywhere they have seen it.' | Affect | – audiences |

| | | |
|---|---|---|
| Al Gore's voice – images of him in front of audiences and as offscreen voice while images are showing what he is speaking about: | | |
| 'The arctic is experiencing vaster ailment. If this were to go sea levels worldwide would go up twenty feet. | Judgement Appreciation | – arctic / inscribed – sea levels |
| 'This is what would happen in Florida. Around Shanghai, around forty million people. The area around Calcutta, sixty million. | Judgement Judgement Judgement | – Florida / invoked – Shanghai / invoked – Calcutta /invoked |
| 'This is Manhattan. The World Trade Centre Memorial would be under water. Think of the impact of a couple of hundred thousand refugees and then imagine a hundred million.' | Judgement Judgement | – World Trade Centre Memorial / invoked – millions of refugees / inscribed |
| Images of the natural disasters: 'We have to act together to solve this global crisis.' | Judgement | – / + solving global crisis / inscribed |
| Written narrator (intertitles in white bock capitals on black background) cut by images of global warming effects: 'NOTHING IS SCARIER THAN THE TRUTH' | Appreciation | – truth |
| Al Gore's offscreen voice on image of the earth taken from space: 'Our ability to live is what is at stake.' | Judgement | + / – ability to live / inscribed |
| Written narrator (intertitle in white block capitals on black background, no sound) 'AN INCONVENIENT TRUTH' | Judgement | + / – truth / inscribed |

Table 1: Analysis of verbal communication, images and sounds.

Verbal communication also contains some expression of Affect, that is, expressions more directly concerned with emotions and with positive and negative emotional responses and dispositions (White, 2005). The very few instances of Affect are either included in promotional features or they are related to the second metaphorical field of MORAL VALUES, a clear example of this is the metaphorical use of the verb 'betray' repeated twice, highlighted by 'auditory bullets' and the rhetorical question which contains it:

DID THE PLANET BETRAY US ...
OR DID WE BETRAY THE PLANET?[10]

The analysis of evaluative language shows that Judgement is by far the most frequent function in the script of the trailer. These expressions are both emotional and overtly linked to moral values which, as a metaphorical field, is rather more difficult to render in images and sound than in words.[11]

Thus, the analysis of the data points towards a tendency for semiotic codes to specialize: the editing of images, sounds (music and sound 'bullet points', voice quality and paralinguistic features), graphic devices (which powerfully emphasize some verbal elements) all enhance the emotional aspects of communication, involving the viewers personally, directly tapping into cognitive metaphors related to the broader field of emotions.

## Conclusions or Contrasting Messages

The inherently intertextual and interdiscursive quality of the trailer blends the discourse of popularization of science and the genres 'documentary', 'journalistic investigation', 'report' and 'public lecture' which directly derive from the film with a heightened emotional quality of involvement derived from genres such as 'action movie', 'disaster movie', 'science-fiction movie'. See for instance the overtly emotional over-emphasized language directly related to the film (underlined by graphics and powerful sound and music)

---

10    Film trailer: <http://www.climatecrisis.net/trailer/>; last accessed 23 May 2008.
11    Not impossible, though: for instance, the images of the New Orleans refugees after hurricane Katrina have a powerful moral impact, but they cannot be analyzed without the verbal contribution of Al Gore's off-screen comment directing the interpretation of these shots: 'This is ruling NOT a political issue so much as a moral *issue*' (transcript of trailer, <www.climatecrisis.net/trailer/> last accessed 23 May 2008).

which is intended as an overtly promotional section in the trailer (belonging in Maier's words to the Explicit Promotional Stages, 2006, p. 105): 'BY FAR THE MOST TERRIFYING FILM YOU WILL EVER SEE', and the end of the trailer:

> NOTHING IS SCARIER
> THAN THE TRUTH
> Our ability to live is what is at stake.
> AN INCONVENIENT TRUTH[12]

The implication is that the trailer promotes the most terrifying of movies because it is not fiction, but rather the 'real' world situation we are all inevitably involved in: 'Our ability to live is what is at stake' (film trailer).[13] This is a rather commonly used device in horror and disaster movies: arousing fear (one of the most basic human emotions: LeDoux, 1998) and is used to attract audiences presenting its cause as 'real' whilst it is actually fictional narrative. In the case of a real threat for humanity, the interdiscursivity of fiction used to enhance the impact of promotional communication on wide audiences might have both the positive effect of attracting the viewer and conveying to large audiences the alarming message put across by the film, as well as the effect of *dejà vu* and fiction.

On the one hand the intertextual quality of the trailer (its images, participants and off screen voices) promotes the topics, ideas and mission of the documentary raising awareness about one of the most relevant global problems for human kind. Whilst, on the other, its interdiscursivity (quality of editing, speed of shots, music, sounds, voice quality of offscreen narrator, intertitles, explicitly promotional features) refers to a complex blend of genres in which the emotional fictional film narratives are not only highlighted, but also promoted as one of the positive characteristics of the documentary. In fact, these narratives are indeed potentially present in the documentary itself, but they are not the focus of the communication.

---

12    <http://www.climatecrisis.net/trailer/> last accessed 23 May 2008.
13    <http://www.climatecrisis.net/trailer/> last accessed 23 May 2008.

The promotional emphasis of the trailer transforms it into an independent multimodal text whose originality resides in its interdiscursivity as an action movie in which the culprits are the viewers. The film, original investigation / report and scientific popularization, is advertised as a thriller, horror movie and disaster movie which provokes powerful emotions by directly involving the viewers into the context of fear and suspense, thus allowing the public to construe identities which are potentially present but not emphasized in the original movie itself. As hypothesized at the beginning, this might have the positive effect of attracting a wide range of viewers, but also of giving the impression of unreality and *dejà vu*, thus defeating the main purpose of the documentary itself which aims at involving the viewers while highlighting the reality of the issue at stake and its danger for humanity.

This study has tried to show possible ways to analyse emotions in multimodal texts. A much wider corpus of texts is needed to investigate how mediated emotions are conveyed multimodally by blending different semiotic codes into what Iedema (2003) calls 'resemiotization'.[14] A corpus of data multimodally analysed would offer insights into the expression of emotions in different genres and genre-chains and their connection to the construal of ideologies and identities.

In conclusion, from the analysis and interpretation of the data, it emerges that the use of emotions in genres which are closely related (in the case-study, one is derived from the other) can convey rather different and even contradictory effects. The promotional function of the trailer is achieved by constructing identities for the film which are emotionally loaded and only represent its most spectacular aspects; by selling it as a disaster movie rather than as a film of denunciation, the trailer tones down its powerful message of global need for awareness and active change while highlighting that of fictionalized impending catastrophe. The distortion of the emotional impact is a widespread and commonly effective device in advertising, the textual category to which the trailer belongs. In this case,

---

14    'Resemiotization is about how meaning making shifts from context to context, from practice to practice, or from one stage of a practice to the next' (Iedema, 2003, p. 41).

however, the emotions evoked and emphasized in the trailer are at times in contrast with the socially committed messages and the call to reality the film promotes.

# References

Baldry, A. (ed.), *Multimodality and Multimediality in the Distance Learning Age*, Campobasso: Palladino, 2000.

——, *A Multimodal Approach to Text Studies in English*, Campobasso: Palladino, 2005.

Baldry, A. and Thibault, P. J. (eds), *Multimodal Transcription and Text Analysis*, London: Equinox, 2006.

Bakhtin, M., *The Dialogical Imagination*, Austin, TX: University of Texas Press, 1981.

Balkin, J. M., *Cultural Software*, New Haven, CT and London: Yale University Press, 1998.

Bauer, M. W. and Gaskell, G. (eds), *Qualitative Researching with Image Text and Sound: A Practical Handbook*, London: Sage, 2000.

Bora, A. and Hausendorf, H. (eds), *Constructing Citizenship*, Amsterdam: John Benjamins, 2003.

Cortese, G. and Duszak, A. (eds), *Identity, Community, Discourse. English in Intercultural Settings*, Bern: Peter Lang, 2005.

Damasio, A. R., 'Emotion in the perspective of an integrated nervous system', *Brain Research Reviews* 26 (2–3), 1998: 83–6.

——, *The Feeling of What Happens: Body, Emotion and the Making of Consciousness*, London: Heinemann, 1999.

——, *Looking for Spinoza: Joy, Sorrow and the Feeling Brain*, Washington, DC: Harvest Books, 2003.

Davies, M. and Ravelli, L. (eds), *Advances in Systemic Linguistics. Recent Theory and Practice*, London: Pinter, 1992.

Dusi, N., 'Le forme del trailer come manipolazione intrasemiotica', in Pezzini, I. (ed.), *Trailer, spot, clip, siti, banner. Le forme brevi della comunicazione audiovisiva*, Rome: Meltemi, 2002.

Fairclough, N. (ed.), *Critical Language Awareness*, London: Longman, 1992.

——, *Language and Power*, 2nd edition, London: Pearson Education, 2001.

——, *Analysing Discourse. Textual Analysis for Social Research*, London: Routledge, 2003.

——, 'Semiosis, ideology and mediation: A dialectal view', in Lassen I., Strunck J. and Vestergard T. (eds), *Mediating Ideology in Text and Image*, Amsterdam: John Benjamins, 2006.

Fairclough, N. and Wodak, R., 'Critical Discourse Analysis', in van Dijk, T. A. (ed.), *Discourse as Social Interaction*, London: Sage, 1997.

Fedeli, D., *Emozioni e successo scolastico*, Rome: Carocci, 2006.

Fries, P. H. and Gregory, M. (eds), *Discourse in Society: Systemic Functional Perspectives. Meaning and Choice in Language: Studies for Michael Halliday*, Norwood, NJ: Ablex, 1995.

Fussell, S. R., 'The Verbal Communication of Emotion: Introduction and Overview', in Fussell, S. R. (ed.), *The Verbal Communication of Emotions. An Interdisciplinary Perspective*, Mahwah, NJ: Lawrence Erlbaum Associates, 2002a.

——, *The Verbal Communication of Emotions. An Interdisciplinary Perspective*, Mahwah, NJ: Lawrence Erlbaum Associates, 2002b.

Goatly, A., *Washing the Brain – Metaphor and Hidden Ideology*, Amsterdam: John Benjamins, 2007.

Hagan, S., 'Visual/Verbal Collaboration in Print', *Written Communication* 24 (1), 2007: 49–83.

Halliday, M. A. K., *Language as Social Semiotics*, London: Edward Arnold, 1978.

——, *An Introduction to Functional Grammar*, 2nd edition, London: Edward Arnold, 1994 (originally 1985).

Halliday, M. A. K. and Matthiessen, C. M. I. M., *An Introduction to Functional Grammar*, 3rd edition, London: Edward Arnold, 2004.

Hasan, R., Matthiessen, C. and Webster, J. (eds), *Continuing Discourse on Language. A Functional Perspective*, London: Continuum, 2005.

Hill, J., 'Crisis of Meaning: personalist language ideology in US media discourse', in Johnson, S. and Ensslin, A. (eds), *Language in the Media*, London: Continuum, 2007.

Hodge, R. and Kress, G., *Language as Ideology*, 2nd edition, London: Routledge, 1993.

Hunston, S. and Thompson, G. (eds), *Evaluation in Text: Authorial Stance and the Construction of Discourse*, Oxford: Oxford University Press, 2000.

Iedema, R., 'Multimodality, resemiotization: extending the analysis of discourse as multi-semiotic practice', *Visual Communication* 2 (1), 2003: 29–57.

Kővecses, Z., *Metaphor and emotion*, Cambridge: Cambridge University Press, 2000.

——, *Metaphor in Culture*, Cambridge: Cambridge University Press, 2005.

Kress, G., *Literacy in the New Media Age*, London: Routledge, 2003.

Kress, G. and van Leeuwen, T., *Multimodal Discourse*, London: Edward Arnold, 2001.

——, 'Colour as a semiotic mode: notes for a grammar of colour', *Visual Communication* 1 (3), 2002: 343–68.

——, *Reading Images. The Grammar of Visual Design*, London: Routledge, 2006 (originally 1996).

Lakoff, G. and Johnson, M., *Metaphors We Live By*, Chicago, IL: Chicago University Press, 1980.

Lassen, I., Strunck, J. and Vestergaard, T. (eds), *Mediating Ideology in Text and Image*, Amsterdam: John Benjamins, 2006.

Ledoux, J., *Il cervello emotivo*, Milan: Baldini and Castoldi, 1998.

Lemke, J. L., 'Discourses in conflict: heteroglossia and text semantics', in Benson, J. D. and Greaves, W. S. (eds), *Systemic Functional Approaches to Discourse*, Norwood, NJ: Ablex, 1988.

Maier, C. D., *The Promotional Genre of Film Trailers: Persuasive Structures in a Multimodal Form*, unpublished doctoral thesis, Aarhus School of Business, University of Aarhus, Denmark, 2006.

Manheim, K., *Ideology and Utopia*, London: Routledge, 1936.

Martin, J. R., *English Text. System and Structure*, Amsterdam: John Benjamins, 1992.

Martin, J. R. and White, P. R. R., *The Language of Evaluation: Appraisal in English*, London: Palgrave, 2005.

Martin, J. R. and Wodak, R. (eds), *Re/reading the past. Critical and functional perspectives on time and value*, Amsterdam: John Benjamins, 2003.

Niemeier, S. and Dirven, R. (eds), *The Language of Emotions*, Amsterdam: John Benjamins, 1997.

Norris, S., *Analysing Multimodal Interaction. A Methodological Framework*, London: Routledge, 2004.

Norris, S. and Jones, R. H. (eds), *Discourse in Action. Introducing Mediated Discourse Analysis*, London: Routledge, 2005.

O'Halloran, K. (ed.), *Multimodal Discourse Analysis. Systemic-Functional Perspectives*, London: Continuum, 2004.

O'Keefe, A., *Investigating Media Discourse*, London: Routledge, 2006.

Pavlenko, A., *Emotions and Multilingualism*, Cambridge: Cambridge University Press, 2005.

Riley, P., 'Epistemic Communities: The Social Knowledge System, Discourse and Identity', in Cortese, G. and Riley, P. (eds), *Domain-specific English. Textual Practices across Communities and Classrooms*, Bern: Peter Lang, 2002.

Scollon, R. and Scollon, S. W., *Discourses in Place: Language in the Material World*, London: Routledge, 2003.

——, *Nexus Analysis. Discourse and the Emerging Internet*, London: Routledge, 2004.

Van Dijk, T. A., *Ideology*, London: Sage, 1998.

——, 'Critical Discourse Analysis' in Schriffrin, D., Tannen, D. and Hamilton, H. E. (eds), *The Handbook of Discourse Analysis*, Oxford: Blackwell, 2001.

——, 'The Discourse–Knowledge Interface' in Weiss, G. and Wodak, R. (eds), *Methods of Critical Discourse Analysis*, London: Sage, 2003.

——, 'Contextual Knowledge Management in Discourse Production. A CDA Perspective', in Wodak, R. and Chilton, P. A. (eds), *New Agenda in CDA*, Amsterdam: Benjamins, 2005.

Van Leeuwen, T., 'Moving English: The Visual Language of Film' in Goodman, S. and Graddol, D. (eds), *Redesigning English: New Texts, New Identities*, London: Routledge, 1996.

——, *Speech, Music, Sound*, London: Macmillan, 1999.

——, *Introducing Social Semiotics*, London: Routledge, 2004.

Van Leeuwen, T. and Machin, D., *Global Media Discourse*, London: Routledge, 2007.

Vasta, N. (ed.), *Forms of Promotion: Texts, Contexts and Cultures*, Bologna: Pàtron Editore, 2006.

Weiss, G. and Wodak, R. (eds), *Critical Discourse Analysis. Theory and Interdisciplinarity*, Basingstoke: Palgrave Macmillan, 2002.

Wenger, E., *Communities of Practice, Learning, Meaning and Identity*, 2nd edition, Cambridge: Cambridge University Press, 1999.

White, P. R. R., 'Appraisal – The Language of Evaluation and Intersubjective Stance', 2005, <www.grammatics.com/appraisal/> last accessed 15 January 2008.

——, 'Evaluative semantics and ideological positioning in journalistic discourse. A new framework for analysis', in Lassen, I., Strunck, J. and Vestergaard, T. (eds), *Mediating Ideology in Text and Image*, Amsterdam: John Benjamins, 2006.

Wierzbicka, A., *Emotions across languages and cultures: Diversity and Universals*, Cambridge: Cambridge University Press, 1999.

Wodak, R., 'What CDA is about – a summary of its history, important concepts and its developments', in Wodak, R. and Meyer, M. (eds), *Methods in Critical Discourse Analysis*, London: Sage, 2001.

——, 'Images in/and news in a globalised world', in Lassen, I., Strunck, J. and Vestergaard, T. (eds), *Mediating Ideology in Text and Image*, Amsterdam: John Benjamins, 2006.

Wodak, R. and Chilton, P. (eds), *New Agenda in CDA*, Amsterdam: Benjamins, 2005.

Wodak, R. and Meyer, M. (eds), *Methods in Critical Discourse Analysis*, London: Sage, 2001.

*Film and trailer*

*An Inconvenient Truth. A Global Warning*, Paramount Classics, directed
   by Davis Guggenheim, USA, 2006.
Website: <www.climatecrisis.net> (last accessed 23 May 2008).
Film trailer: <http://www.climatecrisis.net/trailer/> (last accessed 23
   May 2008).

# Implementing E-Research Environments: The Importance of Trust

TOM DENISON, STEFANIE KETHERS AND NICHOLAS MCPHEE

## Introduction

In this chapter we report on research which explores emerging e-research environments undertaken in Australia as part of the DART (Dataset Acquisition, Accessibility and Annotation e-Research Technologies) project. The research was designed to identify the issues and problems, and the goals, priorities, and constraints that researchers currently experience in their data management practices, in particular as they relate to the use of cyberinfrastructure which, as the US National Science Foundation defines it: '[…] integrates hardware and computing, data and networks, digitally enabled sensors, observatories and experimental facilities and an interoperable suite of software and middleware services and tools' (NSF, 2006, p. 4).

While the overall focus of DART was on technical issues, the research reported here was more concerned with the 'soft' socio-technical issues that contribute to the complexity of the knowledge production lifecycle and which might inhibit researchers in their adoption of such cyberinfrastructure. One of the key findings of this research was that the issue of trust, or distrust, was central to the emotional response to these new e-research environments. Concerns relating to trust were identified in relation to both the technical systems that were designed to manage researchers' data and to the organizational environment in which such systems were maintained. As will be seen in the discussion of the DART project, trust in this instance is strongly related to researchers' perceptions of personal control over their working environments and, more importantly, the conduct of their research. It was concluded that an inability to deal with such issues would lead to

system users attempting to circumvent designers' intentions and to improvise in their use or adaptation of systems and that such behaviour would
severely impact on the take-up of new e-research environments.


## What does trust mean in this context?

There are many approaches to the concept of trust. For example, Kini
and Choobineh (1998) identify three, those of: personality theorists, who
emphasize personality characteristics and readiness to trust; sociologists
and economists, who emphasize the development of trust between individuals and institutions; and social psychologists, who emphasize trust as
an expectation of another in a relationship. They proposed a definition
of trust in information systems as being '[…] an individual's belief in the
competence, dependability, and security of the system under conditions of
risk' (1998, p. 59). A major aspect of this definition is that of risk, in other
words, the fact that the actions of the other party, though important to the
trusting party (trustor), are not completely under the trustor's control. For
a researcher using a data repository, this becomes a more complex issue, as
the trusted party (trustee) is not another human, but a disembodied technical system, together with its equally disembodied organizational context,
which are both outside of the researchers' scope of control.

Investigating trust in the Internet, Dutton and Shepherd (2003) argued
that all technologies are inherently social as a consequence of their design,
use and/or governance. They consider that understanding the social and
institutional frameworks within which complex technologies operate is
essential to the effective use of such technologies, because those understandings influence users' perceptions as to the value and reliability of the
technology in question. Thus, in the context of the Internet and online
services, usage behaviour will be determined by issues such as trust in systems, which must be understood to extend 'to the equipment, people and
techniques essential to the use of online services' (Dutton and Shepherd,
2003, p. 3). In keeping with this approach, they adopted what they con-

sidered to be the common definition of trust, that is, that trust represents a 'confident expectation' (Dutton and Shepherd, 2003, p. 3).

This fits well also within the framework provided by Orlikowski (1999). Her concept of 'technology-in-practice' is a way of looking at the whole systems environment to show that systems implementation will vary not just according to the features and capabilities of the technology being implemented but will be heavily influenced by both the institutional environment within which it is implemented and the attributes and attitudes of those who are to use it. As Orlikowski notes:

> [...] the notion of emergent technology structure, which allows us to frame what users do with technologies as a process of enactment. Thus, rather than emphasising the technology and how actors appropriate its embodied structures, an enacted view emphasises human action and how it enacts emergent structures by interacting with features of the technology at hand. Focusing attention on the emergent structures recognises that while users can and do use technologies as they were designed, they also can and do circumvent built-in ways of using the technology and invent new ways, which may go beyond or even contradict designers' expectations and built-in features (Orlikowski, 1999, p. 6).

Focussing on the issue of trust specifically related to the internet, Volken (2002) argued that trust is central to the diffusion of the Internet because it is a cultural resource which contributes to raising the innovative capacity found within systems by expanding the scope of action while simultaneously reducing transaction, control, monitoring and enforcement costs. Taking a largely economic approach to the concept, he identified two components of trust: rational trust (based on experience) and moral trust (based on values, norms, and attitudes). In his view, systems trust arises when these two components interact, as effective institutional frameworks and governance are required before users will act to adopt systems in innovative ways.

Another definition, used by Kethers et al. (2005) in previous studies, is provided by Mayer et al.:

> The willingness of a party to be vulnerable to the actions of another party based on the expectation that the other will perform a particular action important to the trustor, irrespective of the ability to monitor or control that other party (Mayer and Schoorman, 1995, p. 712).

Building on that definition, Kelton et al. (2008) identify four levels: individual, interpersonal, relational and societal, and argue that all are present to some extent in any given situation. They emphasize the processual nature of creating trust, specifically the importance of experience in determining trustworthiness, whether it be of a person or of a system, and the resulting behaviour of the trustor. In their view, uncertainty, vulnerability, and dependence form preconditions for the presence of trust, while trustworthiness involves: '[...] competence, positive intentions, ethics, and predictability' (Kelton, 2008, p. 367).

In summary, the development of trust requires what Kethers et al. (2005) call 'confidence', and what Luhmann (1988), calls 'system trust' (*Systemvertrauen*), that is, trust not in a person but in the complex network of people, organizational rules, constraints and policies, software systems, technologies and so on, that form systems. As Kelton et al. conclude, 'trust is an attitude composed of two parts: confidence in positive outcomes, and a willingness to modify one's behaviour in expectation of those outcomes' (2008, p. 368).

Before proceeding with a discussion of the research undertaken as part of the DART project, it should be noted that no specific meaning was put to the researchers being interviewed, but the way in which they used the concept of trust would seem to have been broadly in line with this latter definition. As found in the interviews, many scientists are not prepared to be that 'confident' or trusting in the system. This situation is aggravated by the fact that, once a trustor has had one or more negative experiences, the trustor will typically interpret further actions (and, retrospectively, reinterpret past events) of the trusted party in a subjective way that is strongly influenced by her distrust (Kethers et al., 2005). According to Luhmann (1988), distrust therefore has an inherent tendency to become stronger.

# The DART Project

The exploration of the issue of trust within a broad systems environment was undertaken as one of twenty-seven different work pages that made up the DART Project (<http://dart.edu.au/>; Denison et al., 2007). Running from 2006 to 2007, DART was a collaborative project of Monash University, James Cook University and the University of Queensland. Its emphasis was on providing support for the collaborative research process and on adopting a national approach to improving open access to the results of publicly funded research. It was intended to provide 'proof of concept' for a range of software tools to support e-science/e-research and guidance on how best to:

- Collect, capture, and retain large data sets and streams from a range of different sources,
- Deal with the infrastructural issues of scale, sustainability and interoperability between repositories,
- Support deposit into, access to, and annotation by a range of actors, to a set of digital libraries which include publications, datasets, simulations, software and dynamic knowledge representations,
- Assist researchers in dealing with intellectual property issues during the research process, and
- Adopt next-generation methods for research publication, dissemination and access.

These issues were explored in the context of three 'demonstrator projects' in the areas of Bioinformatics and X-ray Crystallography, Climate Research, and Digital History, the latter being chosen specifically to explore the data and information management requirements in the humanities and social sciences.

Overall, DART focused on challenges in lifecycle management, attribution, and provenance of research outputs, but clearly many intellectual property, access, and security issues underpinned the work. What is of interest here are the findings of one DART work package that focused on barriers to researchers using shared data repositories for storing their own data and providing access to others. The main aim of that work package was to explore researchers' attitudes, perceptions, and feelings towards data management in general, and the usage of data repositories in particular.

## DART Field Study

### Study Design

The methodology adopted to explore these non-technical issues was qualitative in nature, based on the grounded theory approach of Glaser and Strauss (1967). As they argued, grounded theory is suited to the emergent nature of field research because grounded theory methods, based on an iterative process of data collection and analysis, provide a unique basis for understanding processes and meanings as ascribed by those working in the area under study.

In order to obtain a realistic and practical view of researchers' attitudes, needs and work practices, data collection included interviews with sixteen researchers from a wide variety of disciplines, and the embedding of information management professionals (IMs) as participant observers in research teams for periods of up to a year. This two-pronged approach had the advantage of collecting observational data about the everyday work of the researchers as well as giving researchers the opportunity to think about and voice their thoughts and ideas on data management.

The interviews were run as semi-structured interviews drawing on a comprehensive list of questions covering research processes, collaboration within and between teams and organizations, data sources and data types, current data management including policies and regulations, and internal and external constraints on data re-use and data management. Where possible, a sample project was captured in an informal process diagram following the Co-MAP process modelling and analysis method, as described by Kethers (2002). The diagram was used to guide questions during the interview, to capture concrete details, and to highlight issues important to the participant researchers. Co-MAP process diagrams focus on the interactions and information flows between the actors in the process. Information flows are captured as arrows between ovals representing the sender and receiver of the information. Graphical symbols represent the media for information flows (for example, a DVD, or a scroll with a seal representing a formal document such as a consent form), and the recipient's perception of the quality of the information flow (e.g. a tortoise for 'too slow', a tick for 'works well', or a stop sign for 'does not occur at all').

In most diagrams, researchers' ideas and 'wish lists' for better data management were also captured, showing the various components such as actors, information flows, information flow contents, media, and quality. As Co-MAP is particularly geared towards locating bottlenecks and weaknesses in a process (Kethers, 2002), the diagrams were also very helpful for capturing issues and problems that interviewees had encountered in their research projects. The symbols used in the final diagram conveyed the quality of information flows as seen by the recipient of the information. These quality symbols thus give an indication of the participant's feelings towards these processes.

To provide an additional perspective, one IM was embedded for approximately one day per week within a research team of Crystallographers and one day per week with a team of Climate Researchers. Another IM was embedded within a Digital History project team for about one day a week. The role of these IMs was to observe researchers' work processes, and in particular their data management processes, to understand their needs, and to provide support where possible.

*Data collection and analysis*

For the interviews, participants (researchers) were selected from a variety of disciplines across the three partner universities. In addition to those working on the three DART demonstrators of Crystallography, Climate Research, and Digital History, researchers from other disciplines were also included in the sample, because they were expected to provide an opportunity to examine the issues within a broader range of contexts. As a result, researchers in fields such as medicine, economics, archaeology, and synchrotron physics, were also interviewed. Interviews were conducted between June and November 2006 and typically lasted between 60 and 90 minutes.

| Discipline | DART involvement | Number of interviewees | Audio recording | Process diagram | University |
|---|---|---|---|---|---|
| Archaeology | No | 1 | Yes | Yes | 2 |
| Climate Research | CR demonstrator | 1 | Yes | Yes | 1 |
| Crystallography | CRY demonstrator | 2 | No | Yes | 3 |
| Crystallography | CRY demonstrator | 1 | No | No | 1 |
| Crystallography | CRY demonstrator | 1 | Yes | Yes | 1 |
| Earth Sciences | No | 1 | Yes | Yes | 2 |
| Economics | No | 1 | Yes | No | 1 |
| Electrical Engineering | No | 1 | Yes | Yes | 2 |
| History | DH demonstrator | 1 | Yes | Yes | 1 |
| Indigenous Studies | DH demonstrator | 1 | Yes | Yes | 2 |
| Indigenous Studies | DH demonstrator | 1 | Yes | Yes | 2 |
| IT | DH demonstrator, project | 1 | Yes | Yes | 1 |
| Medicine | No | 1 | No | No | 1 |
| Physics (synchrotron radiation) | No | 1 | Yes | Yes | 1 |
| Political Science | Project | 1 | Yes | Yes | 3 |
| CR = Climate Research, CRY = Crystallography, DH = Digital History | | | | | |

Table 1: Summarizes the interviews conducted and data collected.

Further information was collected by the DART IMs. The IM embedded in the Digital History project team had been working on a part-time basis for over a year with the Museum Victoria, the Arts Faculty at Monash University, and a rural community on the development of a digital collection based on storytelling as a method of collecting and enriching public history (Pang et al., 2006). As part of her research for a PhD, the project was entrenched in action and ethnographic research principles, with a focus on reflexive design practice. Over the course of a year, many documents were produced with the Museum, the Arts research group, and the community, including a research design document, functional and technical requirements and specifications of the project, the outcomes of usability sessions, design evaluation and re-evaluations, working papers on the history and background of the project, and academic publications. These documents and background were crucial in supplying the IM with explicit and tacit knowledge on the needs of the project team, and the implicit complexities of the various entities in the project function. Similar procedures were adopted by the Crystallography and Climate Research IM, who identified existing workflow processes and data management methods, and documented them. This provided a detailed picture of the way in which the research teams operated, the issues they faced, and the feasibility of potential advances.

As the main focus was on understanding the issues and problems, and the goals, priorities, and constraints that the researchers had in their current data management, both the Information Flow diagrams captured via the Co-MAP process and a series of Strategic Dependency (SD) diagrams, derived from the informal process diagrams, were used heavily in the analysis. The SD model describes the network of relationships among agents (nodes) by specifying dependencies (links) between them (Yu, 1995). A dependency relationship enables the depender to do things that she (the depender) would otherwise not be able to, but also makes her vulnerable if the dependee does not fulfil the dependencies. The model offers four dependency types: goal, task, resource, and softgoal dependencies, which differ according to the degree of freedom they leave for the dependee.

*Issues Raised*

Trust was the main issue, although it was expressed in various ways and was the basis of several problems. Above all trust came out to be connected to data storage and their security. In general, researchers felt their most pressing need was for better storage facilities for their data, but, to be useful, such storage facilities, for example university-run data repositories, had to be reliably available with little or no downtime, and able to provide quick and easy uploading and downloading, as well as regular, automated back-ups of the data. They would also have to provide for different levels of controlled access, including: individual access, workgroup access, access for collaboration partners within and outside the organization, and general access for the public. The need for remote access by the researchers themselves, regardless of where they are in the world, was stressed repeatedly.

The drivers for a high level of data security primarily relate to the need to protect privacy and/or intellectual property. For example, medical researchers, especially when dealing with patient-level data, were concerned not just with controlled access, but also with the location of the data, insisting that data could not transit out of the university. Intellectual property was seen as being especially important if commercial interests were involved.

Before the interviews, it was expected that some disciplines would be much more advanced in their data management practices than others. It was found, however, that with the exception of two researchers from Indigenous Studies, most researchers relied on personal rather than institutional data management facilities. All interviewees stored data locally, on desktops, laptops, external hard drives and (in a few cases) local servers. Most interviewees used a combination of at least two of these storage media – often desktop and personal laptop, or desktop / laptop and external hard drive. Most interviewees also recognized the need for some form of back-up, although the mechanisms varied, from (mostly manual) synchronization of data on multiple devices, to the use of DVDs and CDs (or even email).

These issues are important because preservation and re-use, two of the key concepts that are driving the development of cyberinfrastructure at the policy level, rely on trust, understood in this case as data being securely stored and accessible. There are often also specific organizational requirements, such as those imposed by a university ethics committee or a funding body, as to the long-term retention of data. In many cases, however, there is a lack of awareness of data collection and data management policies and procedures required by the researchers' organizations. So, although there are strong pressures for improved data management procedures, in most cases storage arrangements are based on what is readily available to individual researchers or research teams. In best case scenarios, that may be a faculty or departmental server, but all too often it is the researcher's personal computer.

As noted, information security is essential in this environment, because of the inherent risks in online systems and because without information security there can be no trust in the new environment. This extends also to trust in the organization and people that host the repository. And while establishing an appropriate legal framework for managing information security risks in the new environment is thus central to trust (Lindsay et al., 2007), neither this nor technical security is sufficient if the researchers' perception is that the repository is not maintained well, or that the unit running the repository is not responsive enough to their needs, either now or in the future.

As is demonstrated in this study of e-research environments, the problem of trust includes a range of perceptions and feelings/emotions that are common to researchers and research teams that will need to be dealt with if the use of such environments is to be maximized. The first issue is the emotional impact of the design characteristics of the e-research environment.

Research teams within universities have a strong potential for improvisation, and their motivation to use that potential is so great, that research teams are quite capable of subverting unwanted (e.g. distrusted) systems or of finding ready alternatives that lay more within their control. For example, one scientist told us that he was quite unhappy with having to request a system administrator's help through an (impersonal) helpdesk system, as this resulted in delayed response times and thus more downtime. His personal solution was to email his day's data to himself on a regular basis, thus effectively using the institutional mail server as his backup machine.

This feeling of irritation that may affect researchers in systems designed like that should be taken seriously into account in the literature on the modelling of the implementation of information systems. For example, the work of Heeks (2002) forms an interesting comparison in this context. Heeks modelled the implementation of information systems. Focusing on potential points of failure, he identified 'design-actuality differences' as a powerful contributor to systems failure and suggested that systems which allow 'contingent improvisation' – that is the ability to accommodate design changes and adapt to local needs – are more likely to be successfully implemented. This research has shown that systems which allow for 'contingent improvisation' can also result in problems.

Returning to the definition of trust as proposed by Mayer et al. (1995), a key concept is expressed in the phrase '[...] irrespective of the ability to monitor or control that other party' (Mayer et al., 1995, p. 712). The issues described here instead relate strongly to the need for personal control (or at least that of the research team) over data and work processes. Issues of personal control play a significant role in the data storage issues described above, but they also affect other areas, such as technical data management issues, control over workflows and processes, costing and choice of software, policy as to what data is appropriate to maintain in long-term storage, and, at the end, the emotional sphere and, in particular, trust as well as fear, as we will see below.

Although researchers are aware that centralized facilities are established to meet both their needs and institutional needs, many believe that institutional needs will take priority in centralized services, and that their specific needs will be less important. In this sense they fear a loss of control – that their needs will not be met in an appropriate timeframe and, what is more, that the timeframe will be beyond their control.

There is also a fear and concern that by using centralized services, they may be faced with additional work requirements, leading to forced changes in work processes. For example, it is generally considered that a key feature of centralized data repositories must be the maintenance of appropriate metadata, describing the type of data, its provenance and so on. This is essential for resource discovery, and the accessibility and interpretation of the data. While this is generally recognized as a benefit, researchers fear that the imposition of centralized standards will interfere with their work processes and so cost them time. It is interesting to note that Purdue University in the USA, faced with the same problem, decided to implement a distributed system based on networking and indexing researchers' local storage (Purdue University Libraries, 2006). The problem with that solution, though, is that although it assists resource discovery, it does not improve data management. The tension between good practice and reality was well demonstrated within the field of Crystallography, with one researcher being keen on the concept of centralized storage, while another, despite the risks, preferred the current manual system that simply 'worked'.

In this emotional framework, outlined by a lack of trust and the fear of losing control (and power), personal control is also an issue that arises in terms of costs and funding arrangements. There is a concern that decisions relating to the cost of software and data management solutions, especially when those solutions require that funding be redirected from the acquisition of internal or 'personal' resources to the funding of centrally managed resources, will also reduce researchers scope of action. This can also present practical problems when provision for such funding was not included as part of the research grant and when existing procedures are not perceived as adequate.

Another source of negative feelings, such as stress and annoyance, was the quality of relationship with the organizational IT support department. Many researchers noted that their relationships with their organizational IT support department, whose role includes providing help with data storage, was rather poor. While individual IT support people were praised for their responsiveness and competence, relationships in general were de-personalized, with researchers being obliged to use a helpdesk system to lodge requests for help. Problems, such as server downtime and often lengthy interactions with IT support, were seen as costing time, effort, energy, and ultimately money. Waiting for things to be fixed was considered stressful, annoying, and wasteful. It also emphasized the researchers' lack of control over core aspects of their work processes.

Further evidence of the potential for mistrust was found in connection with the fragmentation of IT services. In the universities studied, IT services are provided by a range of groups. One consequence of this fragmentation is that many researchers seem to develop the idea that existing IT services cannot meet their specific needs. Together with a strong need to feel in personal control, this attitude has led many to make independent arrangements. The fact that, if problems then arose, they received only token support only served to reinforce those perceptions of centralized services.

This seems to be a reasonably common problem, and it is one that must be addressed if new e-research environments are not also to be bypassed. At Monash, this question is being examined by a recently established e-Research Centre, who believes they need to provide a different type of channel between end-user researchers and the capabilities service that exists within the university.

Although one interviewee mentioned the usefulness of a project-wide Wiki in a large, well-organized and strongly structured project with distributed participants from academia and industry, the use of collaborative software tools also appears to be influenced by similar emotional issues. Many interviewees expressed a need for easy-to-use, reliable, robust collaboration tools that work across organizations and behind firewalls, for example annotation software (Indigenous Studies, Digital History), and visualization tools that allow remote sharing of data in real-time (Crystallography), but few had access to these. Some researchers mentioned having tried out different collaboration tools and being disappointed with the reliability (in the case of video conferencing) and speed (in the case of sharing visualizations).

In general, current tools used for remote collaboration are mainly email and phone. This means that these services or devices are more flexible and adapt to manage interpersonal relationships in a satisfactory way. The Earth Sciences researcher reported that he uses skype quite regularly, and the Climate Researcher has used skype with a webcam for weekly meetings with a former colleague now located overseas. Other researchers are aware of skype, but have not yet explored it. The climate researcher also mentioned using a web-based collaboration tool called Basecamp, but stated that one of the project partners did not really use it, which limited the system's usefulness.

Existing data storage arrangements impact negatively on the potential for seamless online collaboration. In the case of local data sharing, researchers will often simply congregate around the same computer and examine the data in single-user mode. Collaboration with remote partners often relies on email (for smaller data sets), and on sending out CDs and DVDs. In the case of the Earth Science researcher, data is often received from and shared with commercial project partners. This can happen through the project's Wiki, or the researcher physically travels to the commercial partner to pick up or deliver the data. Thus, although there is a need to build a collaborative virtual environment for geographically distributed research-ers, only rudimentary collaborative environments exist at the moment, and there remain countless examples of researchers who exchange data by physically exchanging CDs, or by simply using email. A related need is for travelling academics to be able to use what they use at another institution. This is a real need, and solutions are implementable, but because of the cross-institutional dimension to it, solving this issue can be difficult.

But the emotional and relational problems analysed up to now are made worse by another issue. When collaborative tools such as wikis and annotation software are involved, the question of who owns the data created in these environments has the potential to become particularly complex (Lindsay et al., 2007). From the perspective of the barriers to the use of such tools, clarity is once again required. If researchers believe that they are going to lose credit or ownership of their intellectual property they may withhold material, or they may not contribute if they believe that those contributions will not be recognized. Both attitudes weaken the usefulness of these tools, and both can be minimized if an appropriate framework is not only in place but perceived to be in place, and there is trust in the system.

In a situation where there is an established culture and a range of precedents to rely on, it may be possible to create a trusted relationship on the basis of informal understandings. At the present stage of development of e-research infrastructure and collaborative tools, however, neither the culture nor the precedents exist, and formal agreements should be put in place.

It is interesting to note that, although the DART project itself was to make use of collaborative software tools, those tools were rarely used, basically because of these same problems: lack of a secure environment, lack of multi-level access rules, a lack of control over related processes, all of which lead to an inability to incorporate the tools into work practices and a lack of interest on the part of the various DART research teams. In this picture emotions play a vital role.

In line with our findings, there is a growing recognition that despite the technical advances, cyberinfrastructures have not yet managed to fulfil their true potential. Arzberger et al. state that '[...] the barriers to effective data access and sharing are no longer technical, but are institutional and managerial, financial and budgetary, legal and policy, and cultural and behavioural' (Arzberger et al., 2004). David and Spence (2003) argue similarly that technical developments will not be able to transform e-science unless social and technical problem-solving are combined to develop an effective cyberinfrastructure, while Hartswood et al. (2005) note the importance of issues such as '[...] rights, obligations and expectations associated with collegiality, data ownership, confidentiality, IPR, competitive advantage, ethics, and other organizational, personal and professional concerns'.

Achieving the aims of an e-science project therefore not only means solving the technical issues, but also dealing with the organizational and institutional contexts and infrastructures present and, as David (2004) warns, '[...] the socio-institutional elements of a new infrastructure supporting research collaborations ... are every bit as complicated as the hardware and computer software, and, indeed, may prove much harder to devise and implement'. Based on our findings, we would second those statements, but also we would also call attention to the specific role of emotion in the implementation of cyberinfrastructures.

## Conclusion

The DART project set out to develop a prototype e-research system, to satisfy both the practical needs of research teams and the institutional needs of universities in managing their research. The DART work package described in this chapter had as its aim the identification of the barriers to the development and uptake of tools for improved data management procedures by researchers in a collaborative environment. As the research progressed, it became obvious that trust and, perhaps more importantly, perceived trust, were essential attributes of the new e-research environment, and that without it the cyberinfrastructure under development would not achieve its aims. It was also apparent that the concept of trust involved not only trust of individuals or specific system components, but also encompassed system processes and the institutional frameworks within which they operated. The need for trust was closely related to the need for (perceived) control over data storage and access and, in combination with resource costs involved in changing procedures, proved sufficiently strong to force modifications to the overall project.

From the start of the project it was obvious that research teams working in different disciplines would have different requirements. It would also have seemed a reasonable assumption that trust in new systems and the need for a feeling of personal control over processes and data would have been important in the implementation of a new environment. What was not anticipated was the strength of these feelings such as fear and concern and that in this environment that they would have such significant implications.

Our findings regarding the distinctive role played by emotion have implications not only for the features and attributes of new services, but also for the way in which those services are designed. It was concluded that, to the maximum extent possible, new services should be designed with the active participation of the researchers themselves, for example, by undertaking extensive user needs analyses and/or by involving them in an iterative process based on user-centred design principles.

From the perspective of the DART project, this has meant that the tools have had to be developed and to prove themselves also in all respects of the emotional impact before they could attract any interest from researchers. It has only been when that has been done, and researchers can see significant benefits while maintaining control over work processes, that they have been prepared to invest the necessary resources to improve their operating environment. Tangible benefits to researchers, including the immaterial and emotional ones, rather than in principle commitments to professional procedures or non-specific collaboration, are now seen as the only real driver in the uptake of new tools and procedures.

As a footnote, the ARCHER project <www.archer.edu.au>, the successor to DART, took the approach that it would produce a range of tools that, while broadly consistent with the requirements of the universities involved, would be developed in conjunction with researchers, and offered as options through a research portal.

# References

Arzberger, P., Schroeder, P., Beaulieu, A., Bowker, G., Casey, K., Laaksonen, L., Moorman, D., Uhlir, P. and Wouters, P., 'Promoting Access to Public Research Data for Scientific, Economic, and Social Development', *Data Science Journal* (3), 2004: 135–52.

David, P. A., 'Towards a Cyberinfrastructure for Enhanced Scientific Collaboration: Providing its "Soft" Foundations May Be the Hardest Part', Oxford Internet Institute, Research Report No. 4, August 2004. Available at <http://www.oii.ox.ac.uk/resources/publications/RR4.pdf>.

David, P. A. and Spence, M., 'Towards Institutional Infrastructures for e-Science: The Scope of the Challenge', Oxford Internet Institute, Research Report No. 2, September 2003. Available at: <http://www.oii.ox.ac.uk/resources/publications/RR2.pdf>.

Denison, T., Kethers, S., McPhee, N. and Pang, N., *Final Report DART Work Package CR1: Move data from personal data repositories to secure trusted alternatives*, 2007. Available at <http://www.dart.edu.au/work-packages/cr/cr1-final.pdf>.

Dutton, W. H. and Shepherd, A., 'Trust in the Internet: The Social Dynamics of an Experience Technology', Oxford Internet Institute, Research Report No. 3, October 2003. Available at <http://www.worldinternetproject.net/publishedarchive/RR3.pdf>.

Glaser, B. and Strauss, A., *The Discovery of Grounded Theory*, Chicago, IL: Aldine, 1967.

Hartswood, M., Ho, K., Procter, R., Slack, R. and Voss, A., *Etiquettes of Data Sharing in Healthcare and Healthcare Research*, Manchester: First International Conference on e-Social Science, 2005.

Heeks, R., 'Information systems and developing countries: failure, success, and local improvisations', *The Information Society* 18 (2), 2002: 101–12.

Kelton, K., Fleischmann, K. R. and Wallace, W. A., 'Trust in digital information', *Journal of the American Society for Information Science and Technology* 59 (3), 2008: 363–74.

Kethers, S., 'Capturing, Formalising, and Analysing Cooperation Processes: a Case Study', in *Xth European Conf. on Information Systems (ECIS)*, Gdańsk, 2002: 1113–23.

Kethers, S., Gans, G., Schmitz, D. and Sier, D., 'Modelling Trust Relationships in a Healthcare Network: Experiences with the TCD framework', *Proceedings of the XIIIth European Conference on Information Systems (ECIS)*, Regensburg, Germany, 2005.

Kini, A. and Choobineh, J., 'Trust in electronic commerce: definition and theoretical considerations' in *Proceedings of the thirty-first Hawaii International conference on System sciences* (4), Kohala Coast, HI, 1998: 51–61.

Lindsay, D., Menotti, A., Paterson, M. and Chin, A., *Final Report DART Work Package CR6: Legal issues in eResearch*, 2007. Available at <http://www.dart.edu.au/workpackages/cr/cr6-final-080507.pdf>.

Luhmann, N., 'Familiarity, confidence, trust: problems and alternatives', in Gambetta, D. (ed.), *Trust. Making and Breaking Cooperative Relations*, Oxford, 1988.

Mayer, R. D. and Schoorman, J. F., 'An integration model of organizational trust', *The Academy of Management Review* 20 (3), 1995: 709–35.

National Science Foundation (US), Office of Cyberinfrastructure, *NSFs Cyberinfrastructure Vision of 21st Century Discovery*, Version 5.0, 20 January 2006. Available at <http://www.nsf.gov/od/oci/ci_v5.pdf>

Orlikowski, W. J., 'Technologies-in-Practice: An enacted lens for studying technology in organisations', *Information Systems Research* 10 (2), 1999: 167–85.

Pang, N., Schauder, D., Quartly, M. and Dale-Hallett, L., 'User-centred design, e-research, and adaptive capacity in cultural institutions: The case of the women on farms gathering collection', in Khoo, C., Singh, D. and Chaudhry, A. (eds), *Proceedings of the Asia-Pacific Conference on Library & Information Education & Practice (A-LIEP 2006)*, Singapore, 3–6 April 2006: 526–35.

Purdue University Libraries, *Distributed Institutional Repository: White Paper*, 2006. <http://dir.lib.purdue.edu/whitepaper.html>.

Volken, T., 'Elements of trust: the cultural dimension of Internet diffusion revisited', *Electronic Journal of Sociology* 6 (4), 2002. Available at <http://eoe.lac-bac.gc.ca/100/201/300/ejofsociology/2002/vo6no4/volken.html>.

Yu, E. S. K., *Modeling Strategic Relationships for Process Reengineering*, PhD thesis, University of Toronto, Canada. Also published as Technical Report No. DKBS-TR-94-6, 1995.

# Theme 3

# The Emotional Investment Users Put into ICTs

# Emotion, My Mobile, My Identity

JANE VINCENT

## Introduction

By the end of 2008 it is likely that over half of the world's population will have a mobile phone, with some already owning more than one (Short, 2007); however, despite its apparent ubiquity, access to public mobile phone service has only been possible for less than thirty years. Adopted by the majority of populations of many nations (including the UK) for less than half that time it is not surprising that academic research into mobile phone use, especially the development of new theoretical perspectives with regard to social practices, remains relatively new. Furthermore, within this now rapidly growing body of research, there is little literature on the affective aspects of mobile phone usage, in particular with regard to the understanding of emotion and subjectivity. Researchers worldwide continue to record and analyse what people do with their mobile phones and as new populations adopt the capabilities of this versatile computational mobile device it is important to keep building on this rich strand of new learning, particularly with regard to how it might be affecting people's subjectivity.

In this chapter I explore these affective aspects of mobile phones by drawing on a series of my research studies conducted since 2003 with the DWRC[1] in the UK. The foundation for these studies was a three-year research project by DWRC, 'A Socio-Technical Shaping of Multimedia Personal Communications' (Brown et al., 2002; Green et al., 2001; Green,

---

[1]    Digital World Research Centre, Faculty of Arts and Human Sciences, University of Surrey, United Kingdom. www.dwrc.surrey.ac.uk

2002), in which I participated, and by the work of the DWRC Vodafone Scholar (Lasen, 2005). Appendix 1 summarizes seven DWRC studies to show the research topics, methodologies, the size of the samples and reference to the main publications from the studies, that I have used to inform the analysis in this chapter; the last of these studies includes new empirical research not previously published. Each study, some conducted jointly with colleagues, examined the social practices of users of mobile phones in the UK. Whilst the research questions being explored varied they were examined using similar methodological approaches and theoretical perspectives that sought to understand how people used their mobile phones in their day-to-day lives. In the course of the data gathering various themes emerged that were common to each research project and in this chapter I explore two of these, the first of which is about the emotion associated with using the mobile phone and the second is about the ways that the mobile phone is being used in the development and presentation of the self.

In examining the data that supports these topics in this chapter I explore the ways in which these respondents declared their emotion with regard to their mobile phones. Indeed, it would appear that for some this device enabled, or even became imbued with the strong emotion of love and affection between close family members and friends. Although invoked by social interaction between humans I will assert this emotion can also result from interaction only with the mobile phone. Thus I will explore in this chapter how mobile phones may have become incorporated into some peoples' daily lives to the extent that, for many of the respondents in my studies, they have become embedded in the representation of their self.

## Methodological and Theoretical Perspectives

The emphasis for each of the DWRC studies was on attaining new material that captured the lived and personal experiences of the mobile phone users in their own words, and for this reason a qualitative approach to obtaining the data was adopted. Making a diary record of one day in the life of a mobile phone user provided insights but of course their usage could vary for many reasons and the diaries thus provided only a snapshot of their activities within a specified time frame, as indeed did their questionnaires. The material gathered in focus groups and in one to one open ended interviews augmented the diary and questionnaire data with personal anecdotes and experiences as narrated by the mobile phone users. In the final study only open ended interviews were used and these respondents, unconstrained by a diary form or specific dates on which to record their activities, were able to offer up their experiences, providing examples that they felt were most pertinent to them and their feelings about mobile phones. Talking about their use of mobile phones also enabled them to tell of experiences that had shaped why they did things in a particular way; information that could not be so easily obtained by questionnaires or diary records. The interviews and focus groups were transcribed and considered alongside the diary records and questionnaire responses. Emerging themes were identified and in the studies for the UMTS Forum these were validated in peer group reviews with industry and academic experts. The results from each of the studies were used to inform the follow up projects building on the themes already identified as well as exploring new directions.

Literature from the growing body of research that explores mobile phone use also informed these studies but it is of note that in 2002 there was little published academic research, and none on the topic of mobile phones and emotion with regard to the self. Some of the foundations for future publications were laid by the research groups COST 248 (Haddon, 1997), and COST 269 (Kant, 2004; Mante-Meijer and Klamer, 2004), and the seminal publications by Brown et al. (2002) and Katz and Aakhus (2002). Since then published material has grown apace as has the exponential usage and diversity of services available on a mobile phone. During this time the focus for the various DWRC research studies was primarily on the social practices of mobile phone users brought about by communicating with others, although the affective aspects of its use were also explored to some extent as well as playing games, using it as a clock, camera, music player and more.

The topic of this chapter, as described by the title, emotion, my mobile, my identity, is examined from a sociological perspective drawing on the interactionist theories of Goffman (1959) on the presentation of self and of

Hochschild (2003) on emotion work as well as on studies of emotion such as by Kemper (1987), Bendelow and Williams (1998) and Turner (2007). Researching emotion and mobile phones is not without its difficulties due to the multiple ways that emotion and subjectivity can be explored especially when the emphasis is on sociological rather than psychological examination. As we have seen from the introduction to this volume whilst some sociological studies on emotion have been carried out over at least the last one hundred and fifty years it is really only in the last two or three decades that sociologists have begun to specifically examine emotion as a topic on its own. This is exemplified by Shott's assessment of the situation in the late 1970s when emotion was only just emerging as a specific area of research among sociologists, who had until then considered it to be more of a topic for psychology. 'By and large, however, sociologists tend to deal with emotion obliquely and unsystematically, as if reluctant to concede more than slight importance to such a "psychological" factor' (Shott, 1979).

Harré, in his review of the traditional theories of human emotion, suggests that this apparent disinterest in emotion could be attributed to '[...] the predominance, since the seventeenth century of a philosophical conception of emotions as simple, and non-cognitive phenomena, amongst the bodily perturbations' (Harré, 1986, p. 2). Emotions, Harré went on to explain, were 'conceived by philosophers as simple, involuntary and purely affective states' and as such 'not worthy of extensive study in their own right' (Harré, 1986, p. 2). Writing in the late 1990s, Williams and Bendelow also concurred with this view, asserting that a clear body of work on emotion, mostly by American scholars, had only begun to emerge over the previous decade. They suggested that emotions had fallen victim to being considered 'as private, "irrational", inner sensations which may have been tied, histori-cally to women's "dangerous desires and hysterical bodies"' (Williams and Bendelow, 1998, p. xv), and to the quest for scientific knowledge that has placed 'reason rather than emotions as the "indispensable faculty" for the acquisition of human knowledge' (1998, p. xvi).

Examining the data that relates to affective uses of mobile phones is made more challenging when there are so many connotations and descrip-tions of emotions, and variations in approach, a point noted by Kemper in his analysis of the subject. Fundamental in the field of emotions is the

question of how many emotions there are or there can be. The answer proposed here is that the number of possible emotions is limitless (Kemper,
1987, p. 263).

Furthermore, researching emotion demands a lexicon that is commonly understood. However, when even the very word 'emotion' can have
a different sense or meaning for almost every language and culture the
interpretation of the intended meaning of the expressed emotion can be
problematic. As Hochschild (2003), has shown, people express emotion
differently in the context of work and in private life. In addition, one must
ask if there might be some cultural differences in the ways we manage the
emotion of the 'real self', by which she meant the inner, managed heart,
and this is perhaps most manifest in emotion talk and how people express
emotion in words. From an anthropological perspective (Heelas, 1996) it
would appear that there are many different ways that people do this and
from a linguistic perspective there are many interpretations of the expressions of emotion (Davitz, 1969). These studies of emotion in other disciplines highlight just how diverse the research can become and it behoves
the sociologist now to explore emotion specifically in terms of how they
affect the presentation of self.

When analysing the affective social practices of mobile phone users in
these various studies my approach was to examine the experiences recounted
by the respondents and, rather than to look for a particular emotion (such
as hate, fear, anxiety), I explored their actions as well as their expressions
of emotion. In so doing I aimed to avoid imposing any restrictions on the
range and nuance of emotional expressions and feelings that might have
significance in the use of a mobile phone. What is important here is the
intervention of the mobile phone in the social interaction between humans
and in how they are expressing and feeling emotion in ways that include the
device too. For example, the emotion might be expressed bodily such as the
feelings aroused by stroking a mobile phone when you think of a loved one
(Lasen, 2005) or by the emotion invoked by a personalized ring tone for
a special person, or why a particular image is chosen as a wallpaper on the
screen. In my early studies (Vincent and Harper, 2003), some respondents
when asked if they had an emotional attachment to their mobile phone
denied it but then said they could not bear to be without it. Thus although

they had clearly expressed an affective response to their mobile phone they did not think of it in emotional terms, further highlighting that researching emotion is a complex exercise. In addition, as Holland (2007) examines in her paper, researching emotion can be fraught with difficulties that occur when the researcher and her respondent become reactive to each other's emotional response. Discourse on the work of feminist researchers also raises this point. Bondi (2003), writing about empathy and identification, asserts that '[...] when we recognise how someone else feels, we are, in effect, projecting what we have felt onto another'; highlighting the need for caution and perspicacity in the process of conducting and analysing interviews when researching emotion.

In revisiting the data for this chapter, I have examined the references to emotion and the use of emotional epithets as well as exploring the variances in these emotional responses between the studies. This includes statements such as 'I hate ...', or 'I panic ...' and the ways respondents have incorporated the mobile phone into their presentation of self such as 'never being without it'. In the next section of the chapter I will examine and discuss this research material, giving examples of the various experiences of some of the respondents with regard to emotion and their subjectivity.

## The Research

It would appear that for many of the respondents the mobile phone has had a very personal effect on their everyday lives since childhood. In the UK the teenagers who are now entering universities will have grown up not knowing what it was like not to have at least one mobile phone in their household and with children owning them from aged 9 or younger it is now more surprising not to own a mobile phone. Acquiring a mobile phone is a rite of passage for children, especially with regard to the transition from primary to secondary education; 'Because when you go into secondary school they say you know ... I'm growing up. I must have a mobile. I must have my ears pierced' (Chris, age 12, cited in Haddon and Vincent, 2007).

Ling and Yttri (2002) and Harper and Hamill (2005) had similar findings in their respective studies on teenagers use of mobile phones. However, it would appear that although the mobile phone was very important
to the children in my studies they did not use them a great deal. Similar
findings were reported from a study in Finland carried out by Oksman
and Rautiainen (2003). In addition to the monetary constraints their low
usage was attributable to their being at school where mobile phones are not
allowed (although some do use them), or at home where they have access
to other, free to them, communications media such as the internet or the
home phone. Despite this they would often keep their mobile with them
at all times, even sleeping with it just in case they might need to use it, or
to prevent others reading their personal text messages. The mobile phone
was vital for the social life of many children and developing strategies to
minimize their expenditure on it was common. Without a mobile phone
the children felt they might become isolated and as one girl put it when
asked what would happen if she did not have a mobile phone, 'I'd be lonely'
(female focus group, age 14, cited in Vincent, 2004).

The mobile phone for the children aged 11–16 was also demonstrative of their affiliations and friendships through the images and music
they kept on them, the friends they had in their directory and the type of
mobile phone they owned and how they personalized it. Although they
often had to make do with a handed down mobile phone from a sibling or
parent, or a different model than they would have preferred, the ability to
exchange music files and images was often more important than the look
of the device itself. This suggests that displaying ownership of a mobile
phone and sharing what it contains may be more important for some of
these children than actually using it for communications (although tex-
ting was still very important to many), and that the mobile phone is now
seen as a much more versatile and diverse personal computational device.
Exchanging pictures and music is used as a display of their identity and is
demonstrative of their peer group position. The alacrity with which these
children have responded to the new capabilities of their mobile phones has
become an important facet of the presentation of their self within their
peer group such as in the dialogue between these two girls:

ANNABEL: My friend, she forgot her homework. So she looked something up in Google on her phone and wrote the definition down
ALICIA: Wow. Oh, I want Google [...] I'd do my homework on the way to school. (Focus group 3, age 11–12, cited in Haddon and Vincent, 2007).

Indeed, adults and children alike had a special relationship with their mobile phones, returning home for them if they were forgotten, although on occasions they would leave the mobile phone at home on purpose. This apparent contradiction between needing and not wanting to risk losing the mobile phone highlights an emotional paradox in that for some the mobile phone was too valuable to lose (Vincent and Harper, 2003). Some respondents explained how if they were going to a night club and there was nowhere to put the mobile phone or it might be lost when dancing then it would be left at home. The children would not take their mobile to school on a gym day because it was acknowledged that if it was left in the changing rooms it could more easily be stolen. In instances when the mobile phone was intentionally left at home those going to clubs would make advance arrangements for lifts or to text someone when they got home to say they were safe and, as the children knew they were being supervised by an adult and were with their friends anyway, they did not feel dependent on having their mobile phone. Not needing the mobile phone was made possible in these examples by the continued social interaction with friends and responsible adults with the inference being that the mobile phone plays such a significant role in maintaining social contact and personal security when alone that to risk losing it would be unthinkable.

In many instances the emotional pressures surrounding the children and adult respondents' mobile phone use were from other family members or work colleagues rather than from the mobile phone user themselves. Expectations that all people have a mobile phone places an additional burden on those for whom a mobile phone is not their preferred means of communication for making calls. Anthony, interviewed in 2008, only keeps his mobile phone switched on because his friends and, at the time of the interview, his builder call him on it although he prefers to use his home or office phone for outgoing calls. However, as a result of feeling obliged to carry his mobile with him at all times he has started to make use of it in other ways; he uses it in place of his watch and as a phone directory. In these ways Anthony felt he had resisted becoming dependent on his mobile phone but despite this he had found that the role it plays in keeping him socially connected to his friends and family had become vital. Although Anthony does not want to use it as an essential means of communication he has been forced to accept that it has become just that due to his friends' and business contacts' insistence on using it to call him.

In contrast, husband and wife Mike and Mary, also interviewed in 2008 (separately), rely completely on mobile phones in many facets of their daily lives and especially to keep their family safe and in touch, such as when on holidays together with their two teenage children. In their family the mobile phone provides assured contact enabling the more adventurous to visit different locations alone and to find each other if one of the party is separated, particularly when on overseas holidays. On route to their holiday and as they were passing through the airport passport and security Mike discovered they had left one mobile phone in the car. He returned to the car park to collect the missing mobile phone as the inconvenience of doing so was outweighed by the benefits to the family of each having a mobile phone while they were away.

Mike gave another example, however, in relation to his work where the ability to be reached on a mobile phone almost anywhere means that his work colleagues expect him to be accessible twenty-four hours a day, seven days a week:

> You're on the train? Well, that's no reason to stop work because you are on the train and soon it will be the same on planes. At the moment planes are my one little moment of sanctuary where you have to turn the phone off, but not any more (Mike, 2008).[2]

In common with other respondents who expressed similar views, Mike has to 'pretend' to his work colleagues that he does not mind being contacted on his mobile phone when he would prefer to be left alone. This is an example of what Hochschild terms 'emotion work' when the personal

---

2    It had been announced in the press the same week as Mike's interview that mobile phone use would be permitted on some flights from the UK.

feelings and the emotional state of the individual are over-ridden by the needs of those demanding their attention. Although Hochschild's research is predominantly about work situations, this kind of emotional pressure can come from family as well as from work colleagues and may not always be negative.

Simon explains how he is always accessible to his newly graduated children who have recently bought their first homes. The mobile phone is now used as a means of obtaining his advice on many topics especially regarding the house purchase and decorating. He is quite happy to take calls from his daughters whenever they ring and he explains they are simply using the mobile phone as a substitute for the quick questions often posed when they were still living at home: 'There are new things for them to learn and they are learning these via a different medium' (Simon, 2008). This emotionally driven and perhaps occasionally irrational need to ask someone something when it occurs rather than waiting was noted early on in the research. Some respondents reported that '[...] they constantly call their partner/spouse, for example, even when they are in the same house: "I just feel the need to"' (Vincent and Harper, 2003, p. 3). Examples given included, requesting drinks when in the bath or calling someone in another room to come to a meal.

For all age groups and in every study the most common reason given for acquiring the mobile phone in the first place was safety and security. The children wanted a mobile phone so they could call their parents for lifts, or to say where they were. Their parents were happy for them to have a mobile phone so they could call them to check when they did not make contact as expected:

> Mum likes me to have my phone on, especially when I, I'm sort of, out, or away at a friend's house, or something, in case she needs to contact me for any reason, so I don't really turn my phone off that much (Hazel, age 14, cited in Haddon and Vincent, 2007).

In the example of Mike and Mary having mobile phones meant that each family member had more freedom to do what they wanted to do when in unfamiliar settings.

Some adult respondents wanted them in case their car broke down, although several gave rueful accounts of there being no coverage when this did happen and of having to get a lift to somewhere from where they could make a call. It is noteworthy that their priority was to find somewhere that their mobile phone would work over perhaps another solution such as a lift to a garage or a town. One respondent's car had broken down on the way to hospital to give birth to her daughter and neither she nor her husband had a mobile phone. She explained that they had bought one immediately after this experience and that her daughter, now 12, has a mobile phone too so they can keep track of each other. Contrary to Mike and Mary who use mobile phones to reduce the distance and extend the freedom for their children, some parents in an earlier study expressed concern that mobile phones might make their children too dependent on this security link and unable to think for themselves in a crisis: 'You've got to let them live a bit,' said a father and another said: '[...] it takes away your ability to cope [because] you are always relying on the phone' (DWRC, 2002).

Among the older age groups the mobile phone is an adjunct to their personal communications experiences that started before personal computers, and when mobile phones were not commonplace. Nonetheless their mobile phone remains a vital tool in their repertoire of communications media although it has an ordinary and relatively mundane role for many. Although they talked about the mobile as if it was an everyday, functional device they did give examples of using it to store special texts or to keep in close contact with children as they transition from living at home to their newly independent adult lives. One mother talked about how she used the phone to help and manage her everyday life: 'I have three children who all have mobile phones so that I can communicate with them as and when necessary or in emergencies (one has a nut allergy) I now consider them essential' (DWRC, 2002). There were examples in several of the studies similar to this whereby the mother was able to follow her own interests and meet up with friends during the school day when in the past she would have been tied to the home in case of an emergency call from school (Vincent and Harper, 2003; Vincent and Haddon, 2004).

Unlike the children few of the respondents aged over 40 used their mobile phones to listen to music or as a camera except for those who still had young children. James, interviewed in 2007, had two young sons and he recounted a special time he spent with one of them, Jake, as they shared the use of his mobile phone to take pictures and play with it. This mobile phone now had a meaning for him that was much more than just a communications device because of the memories it triggered, and the images that it held of this time spent with his son. During the interview James handled the mobile phone, turning it over and looking at it as he spoke about both his sons but as soon as the conversation moved onto business use he put the mobile down: 'I do associate this [... my mobile phone ...] with Jake because he gets a lot more fun out of it than I do' (James, cited in Vincent, 2007). In a way the mobile phone had become symbolic of his feelings for Jake through their mutual handling and sharing of the device and, for his other son in the use of a photo of him as the wallpaper on the screen. James commented that when he bought a new mobile phone he would give this one to Jake thus continuing the attachment to the device and the shared memories. That Jake might not have the same sentiment for the mobile phone appeared irrelevant; it was the delight his son might get from it that gave James most pleasure. Giving his son the mobile phone conveys perhaps an even stronger emotion for James, that of his love for his son. The cherished memories of times between them imbued in the device are more meaningful to him when played in the hands of his child than if he left the mobile phone on the shelf alongside his previous redundant models.

There were other respondents in the studies for whom the mobile phone had become more than a communications device. In particular, the role of the mobile phone as a directory is a typical use – one that engenders emotion with regard to the potential loss of these numbers. One respondent had lost her entire directory of numbers trying to back it up and now keeps a paper copy of all her mobile numbers just in case. Another, who made a point of manually dialling important numbers to ensure he retained them in his own memory said of his directory: 'I'd really panic if I did lose them, there are about 70 in there' (Anthony, 2008).

It was realized early in the research that mobile phones were used mostly to communicate with people the user already knew (Vincent and Harper, 2003; Vincent and Haddon, 2004), and few numbers, if any, were stored on the mobile phone for people with whom they had no personal contact. Typically these non-personal numbers might be for a car hire company or their doctor. The mobile phone served to help maintain or intensify existing relationships rather than create new and in this way the device itself has become representative of the friends and close family of the mobile phone owner. It is not only the phone directory that is cherished, but text messages, images, music downloads and video clips are all lasting memories of special moments. As we learned from James the information stored on the mobile phone can imbue a special meaning for the user such that it gives comfort when it is physically held and images and texts are revisited to supplant the absent friend or family member. Similarly the children kept hold of their mobile phones because of their attachment to everything stored on it.

Common to all the respondents, whether interviewed individually or in focus groups, was their animated and mostly enthusiastic response to their mobile phone. They wanted to talk about it, to share their experiences and their excitement about what their mobile phone meant to them. Typical of this was one focus group who discussed how they tended to keep their mobile phones with them at all times. Two male respondents said 'It's wherever I am' and 'It never leaves my sight'. One female respondent talked of a 'terrible panic' and of being 'a nervous wreck all evening' when she left her mobile phone at home (DWRC, 2002). The respondents variously showed their interviewer examples of pictures and texts; text messages and calls were answered during the interviews and the device was mostly included in the discussions. When asked why they could not show their mobile phone the few respondents this did apply to expressed discomfort at not having it with them. They had left it in their bag or in an adjacent room because they thought it might not be appropriate to bring it into the interview; others made a point of switching it off out of politeness but then kept it very close to hand.

## Discussion

In these examples from the various DWRC research studies we learn that the mobile phone would appear to have a special place in the lives of the respondents. It is clear that the mobile phone is integral to the maintenance of social lives and family ties, and for some this is brought about by the expectations of others more than by their own desire to use their mobile phone in this way. Fortunati (2005) asserts this continuous interaction between users and their mobile phones may not be entirely beneficial when considered from the perspective of a society that expects more privacy and a slower pace of life than is experienced today. The mobile phone is not merely a communications medium or repository for images, text and voice messages it is as if it has become an extension of the self of its user. It acts as an intra-mediator for the user, functioning as their memory, such as for phone numbers, as a substitute for the presence of a loved one, as a tether between parent and child, and even at times as an intruder when less welcome calls are received. Although we learned the children shared mobile phones they did so out of necessity and in most instances the mobile phone has become an intensely personal and intimate device. Indeed being separated from one's mobile phone is something that the respondents carefully considered and managed with contingencies for the eventuality of actually needing it. There would appear to be a symbiotic link too between the emotion felt about the mobile phone and all that it contains and enables in terms of images, messages and communications and the strong emotion conveyed by it. The sending and receiving of messages containing all kinds of content is not confined to using the mobile phone network or even Bluetooth and infrared, but is now handed on with the showing and sharing of content such as games, personal images, or the showing of unique features for that particular device. The mobile phone has become a third party participant in relationships, a technological link that enables this connectivity but one that also contains tactile links – the very touch of the device triggering strong feelings for some. The relationships are transmuted through the mobile phones in such a way that the device becomes an integral part of the relationship. Families who go on holiday, mothers who go to the gym during school days, parents who are concerned about their children, or a husband and wife who want to be together to share a special moment will use the mobile phone to enable this sense of being together. This unity brought about by the mobile phone extends beyond its role as an individual-to-individual communications device. It seems that the users will use their mobile phone as a substitute for being together with a loved one, be it by looking at shared images or simply by holding the device to evoke the memory of shared moments.

## Conclusion

During the seven-year period of the DWRC research examined in this chapter the various research studies have shown how the social practices of mobile phone users have developed and adapted to incorporate new techno-logical capabilities, such as music playing, file transfer and the camera; the examples given above are but a few of the many contained within these various studies. However, although the capabilities of mobile phones have changed since this research began these changes have augmented rather than replaced the basic communications services of text and talk. This augmentation of the device appears to have intensified the shared and social use of the mobile phone such that it would appear to have a growing presence in the subjectivity of the user. The mobile phone now is not only enabling individual-to-individual communication when apart but through its Bluetooth and Infrared capabilities instant music and picture sharing has become integral to the presentation of the self of many people – especially for the school children discussed above. The type of mobile phone, the pictures they shared and the music they liked were all displayed and showed off to display their identities, as was demonstrated in the various focus groups by the respondents who quickly brought out their mobile phones when talking about themselves.

There were some notable differences between the respondents regarding whom had taken up the new services. The older respondents were less interested in a device with multiple applications whilst the children and adults with young children embraced them with alacrity. However, as the array of new services has grown during the period of my research the affective aspects of the use of the mobile have also developed becoming more intensified and more personal. The mobile phone had always fascinated its users in what it can do in a playful way, as well as to enable communications and give peace of mind. However, it now appears to be so embedded into societal practices that many of the respondents simply assume it to be part of their day-to-day lives, and that of most other people too. This now extends to the more intimate and subjective aspects of their self such as conducting relationships and keeping precious memories. Not having a mobile phone is now something that people cannot easily imagine as it is so omnipresent globally and locally in the lives of individuals and their social groups. Efforts to back up important data such as phone numbers of their friends (some admitting to anxiously copying them onto handwritten sheets of paper) is but one example of the respondents' reliance on their mobile phones as the focus of their social and familial activity. In addition to this is their physical handling of the device whilst they revisit memories through images, text and voice messages or through the shared memories with those whose names are held in their phone directory.

The mobile phone originally intended for use as a mobile adjunct to the fixed line voice communications service enabling people to stay in contact and to give peace of mind when alone has become so much more. It offers a way of showing and sharing identities, of outwardly expressing and displaying one's personal image and of demonstrating affiliations. As shown by the respondents' experiences discussed in this chapter the mobile phone now has multiple roles in the presentation of the self of the user, in the management of emotion work, and most particularly in the role it plays as a third-party participant in the mediation of relationships.

# Appendix : Summary of DWRC/Author's Research on Mobile Phone Users cited in this Paper

| Research topic and date | Methodology and number of respondents | References and comment | |
|---|---|---|---|
| Social shaping of 3G mobile communications technology (2002–3) | Questionnaires, focus groups, interviews: 120 | Vincent and Harper (2003), study for UMTS Forum (included respondents from Germany) | |
| Mobile phone etiquette (2002) | Focus groups: 50 | DWRC (2002), report confidential to client | |
| Social shaping of mobile communications (2003–4) | Focus groups, diaries, interviews: 50 | Vincent and Haddon (2004), study for UMTS Forum | |
| How much do people spend on their mobile phones? (2004) | Questionnaires: 40 | Hamill, Haddon, Vincent, Rickman and Mendoza-Contreras (2004), report confidential to client | |
| 11- to 16-year-olds' use of mobile phones (2004) | Diaries, focus groups, interviews: 105 | Vincent (2004) | Children from the same schools were used for both studies conducted for Vodafone. |
| 11- to 16-year-olds' use of mobile phones and ICT (2006–7) | Diaries, focus groups, interviews: 80 | Haddon and Vincent (2007) | |
| Emotions and Mobile Phones (2007–8) | Open-ended interviews: 30 | Respondents aged 40 and older; study for doctoral thesis (expected 2009) | |

# References

Bendelow, G. and Williams, S. J. (eds), *Emotions and Social Life: Critical Themes and Contemporary Issues*, London: Routledge, 1998.

Bondi, L., 'Empathy and Identification: Conceptual Resources for Feminist Fieldwork', in *ACME: An International E-Journal for Critical Geographies* (2), 2003: 64–76.

Brown, B., Green, N. and Harper, R. (eds), *Wireless World: Social and Interactional Aspects of the Mobile Age*, London: Springer, 2002.

Davitz, J., *The Language of Emotion*, New York, NY and London: Academic Press, 1969.

Denzin, N. K., 'A Note on Emotionality, Self, and Interaction', *The American Journal of Sociology* (89), 1983: 402–9.

DWRC Focus Group Research, *Mobile Phone Etiquette*, University of Surrey, 2002.

Fortunati, L., 'Mobile Telephone and the Presentation of Self', in Ling, R. and Pedersen, P. E. (eds), *Mobile communications. Re-negotiation of the Social Sphere*, London: Springer, 2005: pp. 203–18.

Goffman, E., *The Presentation of Self in Everyday Life*, Middlesex: Penguin Books, 1969 (originally 1959).

Green, N., 'On the Move, Technology Mobility and the Mediation of Social Time and Space', in *The Information Society* (18), 2002: pp. 281–92.

Green, N., Harper, R. H. R., Murtagh, G. and Cooper, G., 'Configuring the Mobile User: Sociological and Industry Views', in *Personal and Ubiquitous Computing* (5), Springer, 2001: 146–56.

Haddon, L. (ed.), *Communications On The Move: The Experience Of Mobile Telephony in the 1990s*, in COST 248 Mobile Workgroup, < www.cost248.org> 1997, retrieved June 2008.

Haddon, L. and Vincent, J., *Growing up with a Mobile Phone – Learning from the Experiences of Some Children in the UK*, DWRC Report for Vodafone, 2007.

Hamill, L., Haddon, L., Vincent, J., Rickman, N. and Mendoza-Contreras, E., *How much can I afford to spend on my mobile*, DWRC Report for Vodafone UK, 2004.

Harper, R. and Hamill, L., 'Kids Will Be Kids; The Role Of Mobiles In Teenage Life', in Hamill, L. and Lasen, A. (eds), *Mobile World Past Present and Future*, London: Springer, 2005: pp. 61–74.

Harré, R., 'A Outline of the Social Constructionist Viewpoint', in Harré R. (ed.), *The Social Construction of Emotions*, Oxford: Blackwell, 1986: pp. 2–14.

Heelas, P., 'Emotion Talk across Cultures', in Harré, R. and Gerrod Parrott, W. (eds), *Emotions: Social, Cultural and Biological Dimensions*, London: Sage, 2000 (originally 1996).

Hochschild, A. R., 'Emotion Work, Feeling Rules, and Social Structure', *The American Journal of Sociology* (85), 1979: 551–75.

——, *The Managed Heart: Commercialization of the Human Feeling*, 20th edition with afterword, Berkeley, CA: University of California, 2003.

Höflich, J. R. and Hartmann, M. (eds), *Mobile Communication in Everyday Life: Ethnographic Views, Observations and Reflections*, Berlin: Frank and Timme, 2006.

Holland, J., 'Emotions and Research', in *International Journal of Social Research Methodology* (10), 2007: 195–209.

Kant, A. (ed.), *User Aspects of ICT Final Report of COST Action 269*, 2004, <www.cost269.org>, retrieved June 2008.

Katz, J. E. and Aakhus, M. (eds), *Perpetual Contact Mobile Communication, Private Talk, Public Performance*, Cambridge: Cambridge University Press, 2002.

Kemper, T. D., 'How Many Emotions Are There? Wedding the Social and the Autonomic Components', *American Journal of Sociology* (93), 1987: 263–89.

Lasen, A., *Understanding Mobile Phone Users and Usage*, Newbury: Vodafone Group R&D, 2005.

Ling, R. and Yttri, B., 'Hyper-coordination via mobile phones in Norway', in Katz, J. E. and Aakhus, M. (eds), *Perpetual Contact Mobile Communication, Private Talk, Public Performance*, Cambridge: Cambridge University Press, 2002.

Mante-Meijer, E. and Klamer, L. (eds), *ICT Capabilities in Action; What People do*, 2004, <www.cost269.org>, retrieved June 2008.

Oksman, V. and Rautiainen, P., 'Extension of the Hand: Children's and Teenagers' Relationship with the Mobile Phone in Finland', in Fortunati, L., Katz, J. and Riccini, R. (eds), *Mediating the Human Body: Technology, Communication, and Fashion*, Mahwah, NJ: Erlbaum, 2003.

Short, M., 'Communications and Mobility', *Appleby Lecture, Institute of Engineering Technology 27 September 2007*, London, <www.iet.tv>, retrieved June 2008.

Shott, S., 'Emotion and Social Life: A Symbolic Interactionist Analysis University of Chicago', *The American Journal of Sociology* (84), 1979: 1317–34.

Turner, J. H., *Human Emotions a Sociological Theory*, London: Routledge, 2007.

Vincent, J., 'Emotion and Mobile Phones', in Nyiri, K. (ed.), *Communications in the 21st Century: Mobile Democracy Essays on Society, Self and Politics*, Vienna: Passagen Verlag, 2003.

——, *'11 16 Mobile' Examining mobile phone and ICT use amongst children aged 11 to 16*, Report for Vodafone, 2004, <www.dwrc.surrey.ac.uk>, accessed May 2008.

——, 'Emotional Attachment to Mobile Phones: An Extraordinary Relationship' in Hamill, L. and Lasen, A. (eds), *Mobile World Past Present and Future*, London: Springer, 2005.

——, 'Emotional Attachment and Mobile Phones', *Knowledge Technology and Policy* (19), Springer, 2006: 29–44.

——, 'Emotion and My Mobile Phone', in *Towards a Philosophy of Telecommunications Convergence*, Conference Proceedings: Hungarian Institute of Philosophy, Budapest, 2007.

Vincent, J. and Haddon, L., *Informing Suppliers about User Behaviours to better prepare them for their 3G/UMTS Customers*, Report 34 for UMTS Forum, 2004, <http://www.umts-forum.org/>.

Vincent, J. and Harper, R., *Social Shaping of UMTS – Preparing the 3G Customer*, Report 26 for UMTS Forum, 2003, <http://www.umts-forum.org>.

Vincent, J. and Harris, L., 'Effective Use of Mobile Communications in e-Government: How do we reach the tipping point?', *Information, Communication and Society* (11), London: Routledge, 2008: 395–413.

# Learning from Emotions Towards ICTs: Boundary Crossing and Barriers in Technology Appropriation

GIUSEPPINA PELLEGRINO

## Introduction

Human emotions are elicited in designing and using technological artefacts, and they affect the shaping of technology itself. On the one hand, especially new technologies are associated with emotional 'labels' in the mass media discourse. On the other hand, technologies themselves elicit emotions of fear, enthusiasm, joy and frustration as they help, or force, to overcome taken for granted boundaries in interaction, communication, space and time. This boundary crossing process, however, can in its turn either enable or constrain appropriation of technologies along sociotechnical trajectories, often allowing barriers to occur. Such a dynamic is illustrated in this chapter with reference to three case studies of technology implementation (two Intranet and one e-learning project) where processes of organizational learning and technology appropriation developed differently as emotions were more or less taken into account.

Drawing from both theoretical reflections on emotional dimensions linked with technology, and empirical evidence collected over comparative studies of technology and organizational learning, the chapter shows how negotiations about technologies – both in practice and in public discourse in the mass media – are emotionally bound. Concluding remarks concern the power of emotional labels and perceptions in sociotechnical construction, and how emotions enable or constrain learning processes about technologies as well as the various social actors involved in their constitution.

## Rationale: Emotions and Technology

When talking about technology and emotions, we are faced with the problem of defining both the terms of the relationship in order to frame their interaction. Depending on how emotions and technology are interpreted, we will have different 'frames of interaction' of these conceptual and practical worlds.

According to the common sense, emotions are often understood in contrast, or opposition, to rational, instrumental aims. Technology is, in turn, associated with 'objective', rationalizing aims, and expected to exclude emotional 'interference' from its planned, inner working. Indeed, according to Oatley (1992), emotions arise when technical plans for action cannot be pursued. If technology, as deriving from the Greek word *téchne*, is at once the result of specific plans for action and the tool to accomplish particular courses of action, it means emotions should not be relevant to it.

However, such a 'black and white' picture of the relationship between emotions and technology is far from giving us back the diverse nuances, complexities and contradictions which characterize the interaction emotions–technology. More and more, in fact, emotions are mediated and expressed through some kind of technological medium; whilst on the other hand, technologies are being shaped, constructed and appropriated through emotions.

Even if cognitive psychology has been searching for an organic framework to describe emotions (cf. Oatley, 1992), there is no general theory able to capture how emotions interact with ICTs and vice versa. Technology itself is subject to different interpretations, more or less oriented to emphasize its technical, social or sociotechnical status (cf. Bijker and Law, 1992; Flichy, 1995; Bijker, 1995; Williams and Edge, 1996).

In this chapter I aim to depict the processes that make emotions more and more directed towards technologies, as well as technologies more and more 'emotional' (to the extent of configuring 'affective' computing). Far from conceiving emotions as 'disturbances' to the process of technological construction, this chapter argues emotional attitudes play a crucial role in the process of appropriation which makes a specific technology, or a set of technologies, perceived as more or less 'familiar', 'useful', 'threatening' or 'indispensable': in other words, part of our everyday life. Appropriation is part of a broader domestication of technology; when a technological arte-fact is appropriated, it is 'owned' by its user, in that it becomes authentic and achieves significance (Silverstone, 1994).

The process of appropriating a technology is always imbued with emotional attitudes and insights, to the extent that neglecting the role emo-tions have in technology construction can lead to misunderstanding and failure, as showed by some of the case studies examined. At a first glance emotions and technology could be seen as separated worlds, having very little to do with each other. Indeed, conceptual overlaps between emotions and technology reveal various levels of interrelation. Emotions seem to belong to the inner psychology of the individual: a world where passions, fantasies and attitudes come from a manifestation of non-rational, often unintentional behaviour. Technology belongs to a material, unambiguous, rationalizing dimension, and so emphasizes instrumentality, synthesis, abstraction, speed and efficiency. Emotions stand for instinct, spontaneity, authenticity, 'naturalness', whereas technology evokes materiality, reason-ing, calculation, artificiality. This polarization most probably summarizes a common-sense, determinist view on technology as well as the prejudice according to which emotion would be an obstacle, even interference, to rational conduct. On the other hand psychologists, biologists and social scientists have shown that emotions are not the outcome of 'rational' action. Instead, they represent states of the mind which prepare for action, sup-porting and inspiring it (Oatley, 1992; Maturana, 2002). We move through emotions as bases for human action, precisely because it is not reason that leads us to action but emotion as 'emotions are corporal dynamics that specify the action dominium we move through' (Maturana, 2002, p. 92). According to Maturana, the intertwining of emotionality and rationality makes possible conversation, which is in its essence based on emotional coordination networks (cf. Villardi and Pellegrino, 2005).

Another point to consider here is Gardner's theory of multiple intel-ligences and especially his definition of 'interpersonal' intelligence (Gard-ner, 1993), that, along with Goleman's analysis (Goleman, 1995), point out emotion is constitutive of action. Emotional intelligence corresponds to the ability to understand one's own and others' emotions. It combines self awareness, managing emotions, being able to motivate themselves and others, handling relationships, showing empathy (Salovey and Mayer, 1990). Emotional intelligence is a type of social intelligence crucial to educational and work processes, as well as to interpersonal relationships and communication. Therefore, emotions have a communicative function in that they lead communication both inside the cognitive system of the individual and in group relationships at home, school and in the workplace (Oatley, 1992; Bar-On and Parker, 2000).

It is not by chance that emotion means 'to move from' or 'to move towards' (from the Latin *e-movere*) and communication comes from the Latin word *moenia* (walls) which has to do with boundaries but also barriers when relating to some otherness (Giaccardi, 2005). Drawing boundaries is a cognitive, social and cultural action, emotionally bound: emotions themselves can be seen as rising from boundaries which restructure and change the context of interaction. The Italian sociologist Gian Primo Cella in his analysis of distinction (Cella, 2006) argues that crossing boundaries means to feel uncertainty as well as anxiety: when boundaries are crossed over, overlapped or renewed, diverse emotions emerge. Passing a boundary means to face otherness, almost a *terra incognita* whose spatial and social 'rules' are unknown and, consequently, need to be re-founded.

In my view emotions and technology are both constituted through complex processes of boundary establishment and boundary crossing which affect communication. Emotions can preserve us from dangers and unforeseen events erecting barriers between ourselves and the current context (e.g. anxiety, fear); emotions can lead us towards new experiences, crossing boundaries which separate us from others and otherness (e.g. joy, enthusiasm). The communicational nature of emotions and of emotional coordination networks (Oatley, 1992; Maturana, 2002) can be interpreted as resulting from a process of boundary crossing. Such a process starts from the nuances of individuality and subjectivity as expressed through non verbal communication and inner feeling. At the same time it emphasizes similarity: emotions are what make us recognize each other as humans. Likewise, technologies can be designed and appropriated so to enable or constrain relationships according to frames of action similar to film scripts

(Akrich, 1992). This boundary crossing process is particularly relevant for Information and Communication Technologies (ICTs) as they aim to mediate, facilitate and transform information exchange and social interaction at a distance. An exemplary case of such a transformation is represented by mobile technologies, namely the mobile phone, which question the taken for granted boundaries between public and private, proximity and distance, closeness and detachment. These technologies provide clear examples of how boundaries in social interaction are themselves on the move (Fortunati, 2001; Katz and Aakhus, 2002).

More generally, boundaries are questioned and emotions elicited or constrained in what can be defined as 'technological mediation of mobility'. Such mediation has to do not only with the mobile phone, but with a more complex set of infrastructures of transport and communication. At the crossroad of such infrastructures different mobilities emerge: information, people, goods, objects, risks are all on the move, configuring hybrid geographies and new patterns of interaction in late modernity landscapes (Urry, 2000; Hannam et al., 2006). These emerging mobilities and the powerful discourse attached to them (Sheller and Urry, 2003), re-shape the way emotions are conveyed, elicited and censored.

Since mobility is linked with the sense of proximity and change (Urry, 2002), it means that emotions attached to physical proximity and face-to-face interaction, so co-extensive to the merging of all the senses and channels of communication, are not simply 'cut off' or 'reduced' through mediated interaction. Rather, the sense of closeness and togetherness, detachment and distance are re-elaborated accordingly, tuned to the situated 'context' enabled by mass and personal media of communication. For example, emotions enabled by television consumption appear partially different than those linked with a personal medium like the mobile phone. As shown in a comparative research across five European countries in the 1990s, classical media (television, radio and telephone) evoke in their users positive emotions whereas the mobile phone is associated with indifference or even frustration (Fortunati, 1998). After more than ten years, it is possible to affirm that the mobile phone elicits diverse emotions, linked with the changing sense of appropriateness and intrusiveness of communication. For example, private conversations on the phone are increasingly 'tolerated' in public space whereas emotions like frustration, anger and joy are conveyed through a portable device. The emerging mobile 'etiquette' and new rules of interaction, therefore, are imbued with emotional attitudes, but also the perception of public and private is transformed (cf. Höflich, 2005).

If communication is like behaviour, that means it is impossible not to communicate and any absence of communication (silence) is communication in itself (Watzlawick et al., 1967), we could say something similar happens with emotions: indifference, or absence of emotions, is emotional itself, in the way it describes an attitude towards something 'other' (being it an object, a person, a behaviour, an event). Given that emotional indifference or indetermination has been considered as a possible attitude towards technology (cf. Fortunati, 1998), it can be argued that indifference prepares for a course of action which considers technology as not relevant to the action itself. Therefore, indifference constitutes a state of mind elicited with reference to technology. Extending to emotions the most renowned Palo Alto axiom (Watzlawick et al., 1967), implies an absolute centrality of them to the establishment and 'success' of interaction: not only human interaction, but also human-to-technology interaction, constitutively since the texture of social interaction is hybrid and always mingled with various technological artefacts (cf. Law, 1997).

## Aim and Method

The core thesis of this contribution starts from the continuity between emotion, communication and technologies (namely ICTs). Such continuity is motivated, as pointed out above, by the communicative nature and function of emotions as a pre-condition of socially competent action (cf. Goleman, 1995). I argue both emotions and technology work out as crossing boundary processes: their interrelation, *a fortiori*, can reinforce or change boundaries, which can become barriers to communication and interaction.

The content of the emotion–technology interrelation is twofold. At both a discursive and a material level, in fact, emotions pervade and orient the process of technology construction and appropriation. I start from the following research questions:

- How are emotions elicited and erased through ICTs?
- How are technologies emotionally bound?
- What can we learn from emotions attached to specific technologies in processes of organizational change?

Answers to these questions are provided by articulating two levels of analysis. In the first part of this chapter the role of emotions is discussed with reference to the multiple public discourses concerning technology: These discourses have the power to induce and arouse feelings of hope, fear, omnipotence, anxiety attached to specific social settings which new technologies would 'enable'. The framework adopted here comes from Social Informatics and especially from Iacono and Kling's analysis of the rise of the Internet as resulting from discursive frames of democracy, progress and freedom (Iacono and Kling, 2001). I will refer in particular to the media (emotions in mobile phone advertisement) and the designer's discourse (ubiquitous and affective computing).

Three general emotional dynamics towards technology are identified as emerging from emotions in technology discourse: anthropomorphization (technology as conversational partner); otherness (technology as 'something out there'); fusion (technology as embodiment of human emotions).

The second part of the chapter presents results of two research projects carried out in three firms implementing respectively two Intranets and one e-learning platform (Villardi and Pellegrino, 2005). Evidence from the fieldwork conducted by using a qualitative ethnographic approach is provided to show how emotional acting (emotional intelligence in Goleman's terms) is constitutive to the success or failure of technology appropriation, and how bounded emotionality and conversation (in Maturana's terms) marked different results in processes of collective learning and appropriation of technology. These different outcomes are depicted in terms of boundaries and/ or barriers. The occurrences of boundaries which can be crossed, instead of barriers which are obstacles to the crossing, are rooted in the mismatch between emotional attitudes towards technology and hyperformalization occurring in organizational contexts. Hyperformalization results from neglecting how pre-existing informal communicational practices would interact with the new artefact. As a consequence, interaction based on new technological artefacts to accomplish core communication activities can easily end with dissatisfaction, perception of failure and frustration. Such an excess of formalization often blinds actors, preventing them from understanding the true potential that emotional learning of technology could provide to their individual and organizational lives (cf. Pellegrino, 2004). Overlooking the strong emotionality dimension in technology implementation (Zorn, 2002), which ranges from feelings of frustration, rage, suspicion and fear to confidence, enthusiasm and collaboration, can make a difference between failure and success, conflict and acceptance, hostility and hospitality towards technology in organizations (cf. Ciborra, 2001). Most organizational research still emphasizes competition/exclusion and restricts cooperation/acceptance only as a means to obtain more competitiveness and better performance. Even if the costs of emotional labour (Hochschild, 1983) have been largely shown, studies in organizational learning and communication keep focusing on emotion and power as means to obtain desired performance. Therefore, organizational knowledge and ICTs are predominantly considered as transferable possessions rather than the result of acceptance and recurring conversation practice.

Final remarks in the chapter will concern a summary of the 'lessons' emerging from considering emotions towards ICTs as crucial domains for designing future frames of technology use and appropriation, as well as boundaries and barriers in organizational processes of technology implementation, where the role of emotions is often overlooked.

# Emotions in technology discourse

## *Fear, hope and omnipotence*

The first level to frame the relationship between emotions and technology is the discursive level that means the public discourse concerning potential uses of technology in current and future social settings. By public discourse, is meant the set of '[...] the discursive practices – the written and spoken public communications – that develop around a new technology. Public discourse is necessary for particular understandings about new technologies to widely circulate' (Iacono and Kling, 2001, p. 110).

Authors of such a discourse comprise very different actors: institutions, designers, experts, the media, professionals, users, writers, artists. However, in what follows I will focus on the media discourse (especially advertisements) and the designers' perspectives on future patterns of interaction (ubiquitous computing, affective computing). These discursive frames represent a meaningful fragment of the wider public discourse built up around technological innovations. They are especially relevant as shaping the context of use, users' expectations and core beliefs associated with technology.

The notion of 'public discourse' I refer to comes from the Social Informatics approach (cf. Kling and Scacchi, 1982; Kling, 1996). In their account of the rise of the Internet and distant forms of work, Iacono and Kling (2001), proposed a rich and detailed theoretical model based on the way meanings of a specific technology are attributed and framed by different groups of actors. These groups defined as computerization movements, whose aims and beliefs are pursued through collective action, constitute an alternative explanation for the rapid emergence and high diffusion of a new technology. These authors point out that 'tales' are built up around key technologies, on the basis of some 'recurrent frames'. Hope, enthusiasm as well as anxiety and fear are frequently elicited and enacted through these tales as narrative emotions, adhering to:

> [...] the conventions of utopian and anti-utopian writing [...] Authors who write within these genres examine certain kinds of social possibilities, and usually extrapolate

beyond contemporary technologies and social relationships. Utopian tales are devised
to stimulate hope in future possibilities as well as actions to devise a self-fulfilling
prophecy. In contrast, anti-utopian tales are devised to stimulate anger at horrible
possibilities and actions to avoid disaster (Kling, 1996, Chapter A, Section II).

Evoking alternative social orders means to provide emotions to relevant
groups of users and to the society as a whole, by configuring new possibilities as well as new threats: hopes and fears, enthusiasm and frustration,
joy and anger. Actors narrating such stories are multiple: governments,
institutions, scientific disciplines, organizations and professional communities, journalists, users, literature and science fiction are all involved,
and the mass media play a crucial role. From this specific perspective, the
media select and set up discursive frames to help their audiences to make
sense of technological innovations and artefacts. Such a process comprises
both providing information about new technologies (e.g. computer and
the Internet) and teaching audiences to use new technologies. Indeed
the media also played the role to promote information literacy in many
European countries, for example through a mediatization of the Net and
a complementary process of Internetization of classical media (Fortunati,
2005a). Enacting emotional patterns helps to produce meaningful frames.
Examples supporting this statement can be drawn from very diverse settings: hopes towards a new, more equal, democratic social order as well as
fears of obsessive control and surveillance are traceable in the debate about
new and old ICTs, from the landline phone to the radio and television, and
moving on to the Internet and surveillance technologies themselves. The
mass media constitute a gatekeeper system through which values, beliefs
and representations of a new technology are filtered and proposed to the
audience of its potential users. Mobile phone and mobile technologies are
not an exception in this regard: through the advertising discourse, based on
emotional frames, technology is associated with individual and collective
experiences of fun, romance, friendship by emphasizing social situations
of interaction instead of information concerning performance, prices, and
services provided by technologies (Aguado and Martinez, 2002).

The advertising discourse, as a specific category of the more general
media discourse, is particularly interesting since it is through this discursive
space that '[...] negotiation between institutional and non institutional

proposals of use and meaning take place. Design tests, commercialization strategies and advertising campaigns become then a relevant part of that negotiation process' (Aguado and Martinez, 2006, p. 3). However, such understandings are not merely linked with the 'good' and 'bad' of technology. They are often translated into dreams and beliefs of technological omnipotence, where artefacts disclose unlimited possibilities to embed knowledge, memory, progress and democracy (e.g. the Internet). Technologies embody contemporary Gods, prospecting solutions to social problems and new patterns for social order.

Pervasiveness of technology in everyday life is the material correspondent of this discursive framing of omnipotence. Ubiquitous computing is a kind of technology that inhabits the everyday life and everyday surfaces: 'Ubiquitous meant not merely in every place, but also in every thing. Ordinary objects [...] would be reconsidered as sites for the sensing and the processing of information [...]' (Greenfield, 2006, p. 11).

Ubiquity, embedded into what Greenfield (2006), labels as 'everyware', is linked with the idea of a pervasive, omnipresent technology, widespread in different places and variety of scales than personal computing (Greenfield, 2006). Such a kind of material process is accompanied by feelings of anxiety, suspicion and fear, linked:

> [...] with the re-discovery of risks of total surveillance and traceability of the 'electronic body'. Permanent connectivity and availability, associated with recording of conversations and messaging under the urgency (and excuse) of international terrorism after September 11th exasperated the emphasis on control and monitoring of communication at a distance (Pellegrino, 2006, p. 142).

*Emotional labels, emotional embodiment:*
*From otherness to anthropomorphization and fusion*

Beside emotions conveyed in technology discourse, there are emotions which are more directly linked with technological artefacts, almost 'attached' to them. These are expressed through emotional labels referred to communication technologies in everyday practice. Emotional labelling signals how emotions are embodied or articulated towards technologies, through relationships ranging from detachment and otherness, to anthropomorphization and fusion.

Detachment and otherness represent one pole of the relationship between technology and emotions. Technology, according to such a relationship, is 'something out there'; an aseptic, cold, black and white tool (see below). The other pole of the relationship is the fusion of emotions inside technology. This represents an emerging type of technology, which attempts to embody emotions into technological artefacts, and even to extend the range and nature of human communication linked with technological artefacts. In fact:

> [...] many strands of academic study and commercial investment have now been woven into a field loosely known as artificial emotion technology or affective computing. This involves developing tools, resources and strategies for use at the point where human emotion meets technological capability (Davitt, 2004, p. 2).

Affective computing aims to reproduce and even transform the potential of human-machine communication, overcoming the limit of emotional un-expressivity on the machine side. Emotion technologies are designed to respond and recognize human emotion; they also aim to stimulate and simulate emotions so that emotion recognition, generation and simulation can be carried through technological systems (Sutch, 2004). In between otherness and fusion, anthropomorphization of technology occurs. Here technologies are represented by users as 'conversational partners' (cf. Zucchermaglio, 1999); the computer is assimilated to a participant in the communicative interaction and work activity. Such participation is often perceived as contradictory and a source of conflict, to the extent that anger, frustration and exasperation, physical attack and swearing at machines are frequent experiences among users of technologies and especially of ICTs (Cenfetelli, 2004).

Even if technologies are often linked with instrumentality and technical skills being dominions of masculinity (Wajcman, 1995), they always trigger some kind of emotion (no matter if this is frustration, disappointment or satisfaction, happiness, involvement and so on). Construction of technology itself, as embodied in social relations and institutions (Bijker,

1995), is led by emotions and passion; sometimes by utopia of possession as in the case of managing knowledge through ICT-based knowledge management systems – and also by desire of establish or reinforce power inside organizations. Therefore, a key question to ask is the following: how is technological mediation imbued with emotions? And how do technologies contribute to enable or constrain emotions? To address such questions, it is important to frame emotions in the context of the diversity between face-to-face and mediated communication. The complexity of cues which characterizes the physical co-presence, in fact, is restricted – indeed, restructured – in mediated communication (cf. Thompson, 1995). If face-to-face interaction and co-presence represent the main experience of others in everyday life and all the other cases represent deviations from such a 'prototype of social interaction' (Berger and Luckmann, 1967), the importance of physical proximity as based on rich, multi-layered and dense conversations (cf. Urry, 2002), would not be totally replaceable by other means of communication. However, if we consider Maturana's concept of emotion, it would be necessary to pay more attention to ways of developing physical and virtual spaces for 'conversations' practice and for developing acceptance within organizations, and especially acceptance towards technology. Also because the 'founding' role played by body-to-body communication is in question and needs to be analysed with reference to new forms of communication (Fortunati, 2005b).

Even if face-to-face encounters keep playing a key-role in establishing and maintaining social interaction and power dynamics, technologies of communication as such can foster or constrain processes of learning based on emotions. An example is provided by computer-mediated-communication (CMC) whose performativity also comes from enabling and fostering different ways to support emotional exchange, for example emoticons resembling non-verbal communication.

The intertwining of emotionality and rationality is crucial to technology construction and alignment of Information Systems infrastructure. As argued by Ciborra (2000, pp. 30–2), technologies need to be cultivated and taken into account through a process of care. Care means familiarity, intimacy and continuous commitment with reference to the process of technology implementation, design and use. Technology, being both fragile and ambiguous, is in a state of flux in organizations, and its accept-ance takes the form of hospitality (which can, in turn, become hostility) as: '[...] one of the oldest arts of mankind: hosting a stranger' (Ciborra, 2001, p. 30).

To summarize, communication and social interaction have emotional bases which cannot be deleted and which are constitutive of organizing as texture (Cooper and Fox, 1990) and learning as process, based on knowing-in-practice (Gherardi, 2000, 2001). Technology plays an ambivalent role in supporting such processes, as it will be illustrated through the case studies presented in the following section.

## Technology Implementation, Organizational Change and Emotions: Results from Three Case Studies

Organizational settings represent an interesting observatory to focus on the role of emotions and the relationship between ICTs and emotions. Organizations often lead emotions towards specific patterns, exploiting them to the extent of configuring an 'emotional labour' as mentioned earlier (Hochschild, 1983). Furthermore, looking at the interrelation emotions-ICTs in organizational contexts, it is possible to frame emotions towards technology appropriation as affecting processes of organizational change and learning. Results of fieldwork research carried out in three companies located in Brazil, Italy and Great Britain (Villardi and Pellegrino, 2005),[1] will be used to explicate how emotions can make a difference in the per-ceived failure or success of technology implementation inside specific organizational contexts. Furthermore, these results show how processes of organizational learning and change (Dutta and Crossan, 2003), must be driven by emotionality, care and commitment in order to be effective.

---

1    The research was carried out in the Brazilian firm by Beatriz Quiroz Villardi, in the Italian and the British firm by Giuseppina Pellegrino.

The fieldwork was carried out using a qualitative, socio-constructivist and reflexive methodology (Alvesson and Sköldberg, 2001). Research in the Brazilian telecom firm (Alfa) as well as the Italian and the British firm (Beta and Gamma) used the ethnographic approach, combining complete participant observation, semi-structured interviews, informal ethnographic interviews and documentation analysis. Complete participant observation was configured because during the research period the researchers worked within the organizations, workers and managers knew that they would be the object of this research, and accepted the researchers as members of their group (cf. Villardi and Pellegrino, 2005).

The cases present differences and similarities. All of the firms operated in the information technology (IT) sector (telecommunications; integration and management of Information Systems). Each of them wanted to start a process of organizational change by implementing and adopting some kind of technology. Therefore, the process of organizational change was meant to be 'driven' (or at least supported) through a technological-based project (e-learning process at Alfa; the Intranet as platform for knowledge management at Gamma).

Alfa wanted to develop its managers through a two year e-learning project conjointly developed by its own Human Resources staff and a Business School. First results of the project showed a new attitude towards e-learning for the 120 participating managers of the IT firm and innovative educational services for the Business school. Emerging power dynamics and intense emotions within coordination staff revealed interactive relationships between individual cognition and action and generated individual tension when, for example, course content was analysed and changes were made to stimulate students e-learning. Tutors realized that the new medium would require them to update their face-to-face courses while teachers tried to posit that they were already updated. However, acceptance seemed to prevail within their interactions and permanent discussion as a practice allowed negative emotions to change and new conversations to occur recurrently.

It could be said that the e-learning project enacted a more comprehensive process of organizational learning: both were based on an emotional approach towards technology and organizational practice, so that the new technology became part of networks of conversations and acceptance in Maturana's terms. In the Brazilian case (Alpha), an engaged organizational learning process, inspired by inclusive logic and dialogue, made learners and actors aware of the importance of emotional engagement and passionate knowledge in implementing the e-learning platform. An emotional learning of technology was enacted.

Both Beta and Gamma operated in the information systems sector of IT, and wanted to build up and enhance processes of organizational change by implementing Intranet technology. In particular the Italian firm's Intranet project aimed at developing this technology as a 'building site' to familiarize the company staff with web technologies. Such a project was born in the context of a contradictory intra-organizational relationship with the firm's parent company. Such a context heavily affected emotions towards the Intranet, and especially a pervasive sense of 'otherness' towards it. On other hand, the British company wanted to improve processes of knowledge sharing and transfer in order to reinforce identity and belonging to the organization. Such a strategy was pursued by buying and customizing an Intranet-based knowledge management system.

In both the Italian and British case lack of care, commitment and hospitality towards technology manifested themselves through a sense of frustration, disappointment and failure within the process of technology implementation and organizational change. I will now focus on these two cases in order to analyse how emotions can be elicited and erased within a process of technology implementation and appropriation.


## Eliciting and Erasing Emotions through Technology

In both the Italian and the British companies the principle of acceptance (Maturana, 2002), was often overlooked in constructing the Intranet as new technology, thinking of it as abstracted and detached from the context and the organizational everyday life of the companies' members. Therefore, technology was used to erase emotions from the organizational texture, through a process of censoring or constraining passion. In particular in the British case, two competing dimensions of knowledge emerged: one of knowledge as linked with belonging, identity, feelings of proximity and integration inside the organizational setting; the other of knowledge as commodity searchable and accessible through the Intranet system irrespective of time, space and situated practice. The system was bought from an external software provider. The Intranet, named 'the Compass', was used as a tool to orient people towards organizational knowledge and was designed around company consultants working at client sites. However, inter-organizational and socio-technical constraints (e.g. clients preventing consultants from accessing the Compass from their sites, off-line communication, and ambiguity in defining search engine mechanisms and knowledge itself) meant that the Compass was not used as intended.

Controlled communities were implemented in both the companies in order to enable knowledge sharing via technology, and failure of such an attempt could be attributed to marginalization of emotions, passions and relationships of spontaneous engagement on which learning communities are founded (Lave and Wenger, 1991). The way distance was created and overcome through Intranet-based communities frustrated the process of use and appropriation. Finding out what each other was doing was reduced to typing a question onto a machine, or enforcing communication through artificial and abstract 'groups' which had no correspondence with the organizational everyday practice.

On the other hand, in both the settings there was an emotional investment in the technology. For both companies it was aimed at eliciting or supporting some kind of emotional exchange, e.g. the Italian company was using the Intranet as a self-presentation device and the British company wanted to make consultants feel less alone and isolated at the client site. In this case the Intranet was meant to be like a '[...] community, like an online chat to make consultants feel they are part of the company [...] Actually we built a library rather than a social centre. In a library you can get books, but in a social centre you can talk to people, exchange information and so on [...]' (managing director, British firm).

Integration of the Intranet with older media was constantly observed on the field. As it emerged from some of the interviews carried out, some media were felt, perceived and used as more suitable to different circumstances than others. Furthermore, different degrees of effectiveness and formality were associated with them. For example e-mail was described very frequently as a formal and cold tool erasing emotions or making their communication very difficult, as in the following interview excerpts:

> E-mails are very cold, black and white [...] opened to misinterpretation which is quite a big thing since something black and white doesn't necessarily tell you that what you want to, what you expect because when you come across something you don't know if the person who wrote it was in a rush or just nasty with you while generally telephone or conversation get less misinterpreted than e-mail (Account Manager, British firm).

Asking people what the Intranet and other communication media did not mediate, a recurrent answer was 'emotions': 'Technology is an aseptic tool. It is very difficult to express emotions if you aren't a very good writer' (team co-ordinator, Italian firm).

The Beta and Gamma cases in Italy and the UK showed just how lack of commitment and care towards technology, as well as censorship of current practice and previous emotional networks can bring about failure, frustration and misuse of technology. All these results, however, are imbued with emotional ties. They are emotionally bound. Is there anything we could learn from this?

## Conclusion

This contribution has identified two main profiles of the relationship between emotions and technologies of communication (ICTs), even emphasizing that such a relationship is constituted through complex and multi-dimensional dynamics.

The first analytical level is the discursive one. Narrative emotions attached, especially by the mass media, to new technologies (in particular those which become new communication media) constitute a frame to convey emotional images and meanings, to lead processes of technology appropriation and create expectations in groups of users and audiences.

Public discourse on technology circumscribes technology as imbued with emotions of fear, joy, enthusiasm, anger and frustration. These frames constructed in multiple public discourses are then appropriated and translated through micro processes inside organizational settings. These microprocesses represent the second analytical level.

Since both emotions and technologies can be characterized as resulting from communication trials aimed at crossing boundaries previously established, I have argued that such boundaries can turn into barriers to communication itself. When going from the discursive level to the organizational level, that means considering situated contexts where specific sets of ICTs are appropriated and domesticated, such a transformation of boundaries into barriers becomes more evident. Emotions can help to overcome boundaries if an emotional learning process is enacted as in the Alfa case, that means if emotional intelligence and emotions towards technology are not censored inside the organization. When technology is used to erase emotions (cf. the Beta and Gamma cases), boundaries can turn into barriers to learning, conversations and acceptance, hospitality and care into hostility towards the technology.

Depending on how emotions are elicited or erased to interpret and use technology, technology itself and communicational routines can be more or less inclusive of users' needs: technologies reveal a potential role of enacting boundaries or barriers, when the intertwining of emotionality and rationality – the continuity between emotion, technology and communication – is not taken into consideration. Overlooking emotions as a crucial dimension of sociotechnical action risks the transformation of boundaries into barriers to technology appropriation. Frustration, anger, feelings of failure and misuse then emerge. Technology turns into a potential source of risks and otherness, something 'out there', external and not integrated in processes of organizational change.

Therefore, emotions are not an 'additional' dimension to ICTs. Instead, they are totally co-extensive to them at both the symbolic level (technology discourses/emotional labelling) and the material level (organizational practices of technology implementation; design for future interaction, e.g. affective computing).

Paying attention to emotional boundaries and barriers occurring in ICT implementation and appropriation means to enact learning processes with reference to current and future technological development, not only focusing on the performance and productivity of mediated communication, but also considering emotional appropriateness as crucial to construction of sociotechnical systems.

In the end, nurturing emotional learning of technology instead of censoring emotional intelligence and emotions towards technology can make a fundamental difference for both individual and collective actors. Such a difference must take into account that the technological mediation of communication reproduces but also transforms the enthusiasm, seduction and anxiety of the physical proximity. In this sense, emotions in and of mediated communication do not simply replicate the face-to-face communication emotional patterns. Indeed, technology continues to exert seduction and repulsion, to move people towards new horizons and to inspire passionate action.

# References

Aguado, J. M. and Martinez, I. M., 'Performing Mobile Experiences: The Role of Media Discourses in the Appropriation of Mobile Phone Technologies', paper presented at the XVI World Congress of Sociology, Durban, South Africa: University of Kwa-Zulu Natal, 23–9 July 2006.

Akrich, M., 'The Description of Technical Objects', in Bijker, W. E. and Law, J. (eds), *Shaping Technology/Building Society. Studies in Sociotechnical Change*, Cambridge, MA: MIT Press, 1992.

Alvesson, M. and Sköldberg, K., *Reflexive methodology. New vistas for qualitative research*, London: Sage Publications, 2001.

Bar-On, R. and Parker, J. D. A., *The Handbook of Emotional Intelligence. Theory, Development, Assessment, and Application at Home, School and in the Workplace*, San Francisco, CA: Jossey-Bass, 2000.

Berger, P. L. and Luckmann, T., *The Social Construction of Reality*, New York: Doubleday, 1967.

Bijker, W. E., *Of Bicycles, Bakelites and Bulbs*, Cambridge, MA: MIT Press, 1995.

Bijker, W. E. and Law, J. (eds), *Shaping Technology/Building Society. Studies in Sociotechnical Change*, Cambridge, MA: MIT Press, 1992.

Cella, G. P., *Tracciare confini. Realtà e metafore della distinzione*, Bologna: Il Mulino, 2006.

Cenfetelli R. D., *Getting in touch with our feelings towards technology*, Academy of Management Best Conference Paper, 2004.

Ciborra, C. U., *From Control to Drift. The Dynamics of Corporate Information Infrastructures*, Oxford: Oxford University Press, 2001.

Cooper, R. and Fox, S., 'The "Texture" of Organizing', *Journal of Management Studies* 27 (6), 1990: 575–604.

Davitt, J., 'Emotion technology', 2004, article retrieved at <www. futurelab.org.uk>.

Dutta, D. K. and Crosssan, M. M., 'Understanding change: what can we "learn" from organizational learning?', in Simm, D., Easterby-Smith, M. and Simpson, B. (eds), *5th Organizational learning and knowledge International Conference Proceedings*, Lancaster University, 30 May–2 June 2003.

Flichy, P., *L'innovazione tecnologica*, Milan: Feltrinelli, 1995.

Fortunati, L. (ed.), *Telecomunicando in Europa*, Milan: Angeli, 1998.

——, 'The Mobile Phone: An identity on the move', *Personal and Ubiquitous Computing* (5), 2001: 85–98.

——, 'The mediatization of the net and the internetization of the mass media', *Gazette. The International Journal for Communication Studies* 67 (1), 2005a: 27–44.

——, 'Is body-to-body communication still the prototype?', *The Information Society* (21), 2005b: 53–61.

Gardner, H., *Multiple intelligences: The theory in practice*, New York: Basic Books, 1993.

Gherardi, S., 'Practice-based theorizing on learning and knowing in organizations', *Organization* 7 (2), 2000: 211–23.

——, 'From Organizational Learning to Practice-based Knowing', *Human Relations* 54 (1), 2001: 131–9.

Giaccardi, C., *La comunicazione interculturale*, Bologna: Il Mulino, 2005.

Goleman, D., *Emotional Intelligence: Why It Can Matter More than IQ?* New York: Guilford Press, 1995.

Greenfield, A., *Everyware. The Dawning Age of Ubiquitous Computing*, Berkeley, CA: New Riders, 2006.

Hannam, K., Sheller, M. and Urry, J., 'Editorial: Mobilities, Immobilities and Moorings', *Mobilities* 1 (1), 2006: 1–22.

Hochschild, A. R., *The Managed Heart: The commercialization of Human Feeling*, Berkeley, CA: UCP, 1983.

Hoflich, J., 'The mobile phone and the dynamic between private and public communication: Results of an international exploratory study', in Glotz, P., Bertschi, S. and Locke, C. (eds), *Thumb Culture: The Meaning of Mobile Phone for Society*, Bielefield: Verlag, 2005.

Iacono, S. and Kling, R., 'Computerization Movements. The Rise of the Internet and Distant Forms of Work', in Yates, J. and Van Maanen, J. (eds), *Information Technology and Organizational Transformation. History, Rhetoric and Practice*, Thousand Oaks, CA: Sage Publications, 2001.

Katz, J. E. and Aakhus, M. (eds), *Perpetual contact. Mobile communication, private talk, public performance*, Cambridge: Cambridge University Press, 2002.

Kling, R. (ed.), *Computerization and Controversy. Value Conflicts and Social Choices*, 2nd edition, San Diego, CA: Academic Press, 1996.

Kling, R. and Scacchi, W., 'The Web of Computing: Computing Technology as Social Organization', *Advances in Computers* (21), 1982: 1–90.

Lave, J. and Wenger, E., *Situated learning: legitimate peripheral participation*, Cambridge: Cambridge University Press, 1991.

Law, J., *Heterogeneities*, Lancaster: The Centre for Science Studies, Lancaster University, 1997, 20 May 2006, <http://www.comp.lancs.ac.uk/sociology/papers/Law-Heterogeneities.pdf>.

Maturana, H., *Emoções e Linguagem na educação e na política*, Belo Horizonte, Minas Gerais: UFMG, 2002.

Oatley, K., *Best Laid Schemes. The Psychology of Emotions*, Cambridge: Cambridge University Press, 1992.

Pellegrino, G., 'Hypermediatization and inertia: patterns of technological appropriation between reproduction and innovation', paper presented at the 20th EGOS Colloquium, Ljubljana, Slovenia: University of Ljubljana, 3–5 July 2004.

——, 'Ubiquity and Pervasivity: On the Technological Mediation of (Mobile) Everyday Life', in Berleur, J., Nurminen, M. I. and Impagliazzo, J. (eds), *Social Informatics: An Information Society for all? In remembrance of Rob Kling*, Boston: Springer-IFIP International Federation for Information Processing (223), 2006: pp. 133–44.

Salovey, P. and Mayer, J. D., 'Emotional intelligence', *Imagination, Cognition, and Personality* (9), 1990: 185–211.

Sheller, M. and Urry, J., 'The New Mobilities Paradigm', *Environment and Planning* 38 (2), 2006: 207–26.

Silverstone, R., 'Domesticating the Revolution: Information and Communication Technologies and Everyday Life', in Mansell, R. (ed.), *Information, Control and Technical Change*, London: ASLIB, 1994: pp. 221–33.

Sutch, D., 'Emotion technology', focus document, 2004, retrieved at <www.futurelab.org.uk>

Thompson, J. B., *The Media and Modernity. A Social Theory of the Media*, Cambridge: Polity Press, 1995.

Urry, J., *Sociology beyond Society. Mobilities for the 21st Century*, London: Routledge, 2000.

—— 'Mobility and Proximity', *Sociology* 36 (2), 2002: 255–74, <www.its.leeds.ac.uk/projects/mobilenetworks/>, retrieved October 2005.

Villardi Quiroz, B. and Pellegrino, G., 'Face-to-face and distant learning as emo-rational microprocesses: Understanding change through collective learning from within', in Gherardi, S. and Nicolini, D. (eds), *The passion for learning and knowing. Proceedings of the 6th International Conference on Organizational Learning and Knowledge*, Trento: University of Trento e-books, 2005.

Wajcman, J., 'Feminist theories of technology', in Jasanoff, S., Markkle, J. E.,
    Petersen, J. C. and Pinch, T. (eds), *Handbook of Science and Technology
    Studies*, Thousand Oaks, CA: Sage Publications, 1995.
Watzlawick, P., Beavin, J. and Jackson, D., *Pragmatics of Human Com-
    munication*, New York, NY: W. W. Norton and Co., 1967.
Williams, R. and Edge, D., 'The Social Shaping of Technology', *Research
    Policy* (25), 1996: 856–99.
Zorn, T., 'The Emotionality of Information and Communication
    Technology Implementation', *Journal of Communication Management*
    7 (2), 2002: 160–71.
Zucchermaglio, C., 'Etnografia al lavoro: studio sulle pratiche lavorative',
    *Studi organizzativi – Nuova Serie* (1), 1999: 167–87.

# Contributors

NAOMI BARON (nbaron@american.edu) is Professor of Linguistics in the Department of Language and Foreign Studies at American University in Washington, DC. A former Guggenheim and Fulbright Fellow, Baron is the author of seven books covering topics such as computer languages and language in an online and mobile world. Baron's present research is a cross-cultural comparison of mobile phone use by university students in Sweden, the US, Italy and Japan.

MARIA BORTOLUZZI (maria.bortoluzzi@uniud.it) is a lecturer in the Faculty of Education with the University of Udine, Italy. She holds a Master's in Linguistics (Lancaster) and a PhD in Applied Linguistics (Edinburgh). Her research interests and publications are in the fields of Critical Discourse Analysis, multimodality, teaching and learning English as a foreign language. She has numerous publications within the fields of linguistics and applied linguistics.

TOM DENISON (Tom.Denison@infotech.monash.edu.au) is a research associate with the Centre for Community Networking Research in the Faculty of Information Technology at Monash University. Having consulted widely in Australia and Vietnam, his research interests relate to the provision of online services within the frameworks of social and community informatics. His current projects include: a cross-cultural study of the drivers for, and barriers to, the adoption of web-based technologies by non-profit organizations in Australia and Italy; and a study of the role of social networks and the use of information and communications technology for social cohesion among Chinese and Italian communities in Melbourne, Australia, and the Chinese in Prato, Italy.

LEOPOLDINA FORTUNATI (fortunati.deluca@tin.it) is Professor of the Sociology of Communication at the Faculty of Education University of Udine, Italy. She has conducted extensive research in the field of gender studies, cultural processes and communication and information technologies. Her work has been published in eleven languages and she is a member of several European networks including COST 298.

JOACHIM HÖFLICH (joachim.hoeflich@uni-erfurt.de) is Professor for Communication Science with focus on Media Integration at the University of Erfurt, Germany. His research is on media use and media effect; 'new' communication technologies and change of mediation cultures; theory of (technically) mediated (interpersonal) communication, intercultural comparative media communication and mobile communication.

STEFANIE KETHERS has contributed to numerous interdisciplinary projects focussing on her main research interest, supporting human co-operation. She has published more than twenty scientific papers and has reviewed papers for several international conferences and workshops. Stefanie received her doctorate degree from RWTH Aachen, Germany, in 2000 with a thesis on modelling and analysing co-operative processes. She then continued her research at CSIRO in Melbourne, Australia. In 2006, she worked for Monash University in Melbourne, Australia, as a senior research fellow.

NICHOLAS MCPHEE has recently obtained his Master's Degree in Information Management and Systems from Monash University. With a background in information technology, he has been involved in a number of collaborative research projects relating to interface design and the development of software tools to support e-research in academic environments, and the broader social issues surrounding the use of such tools.

GIUSEPPINA PELLEGRINO (gpellegrinous@yahoo.com) is lecturer and researcher at the University of Calabria. She received a PhD in Science, Technology and Society from the same university in 2004 and currently teaches Social Communication at the Faculty of Political Sciences. She was visiting fellow at the University of Edinburgh Research Centre for Social Sciences (2001), the Lancaster University Centre for Mobilities Research (2007) and the Institute for Advanced Studies in Science, Technology and Society, Graz (2008).

SATOMI SUGIYAMA (ssugiyama@fc.edu), PhD in Communication, Rutgers University, is an assistant professor at Franklin College Switzerland. She has been conducting research in Europe, the USA and Japan on communication technology, fashion and communication, and intercultural communication. She has published on mobiles as fashion statements and the co-creation of mobile communication and public meaning.

JANE VINCENT (jane.vincent@surrey.ac.uk) joined the University of Surrey's Digital World Research Centre as a Research Fellow in 2002 after spending over twenty years in the UK mobile communications industry. Her research interests are in the user behaviours associated with mobile communications and in emotions and mobile phones, which is the topic of her doctoral research and on which she has published numerous articles. She is a member of the European network www.cost298.org.

# Index

# Interdisciplinary Communication Studies

## Series Editor: Professor Colin B. Grant

This series publishes research (monographs and edited volumes) of an international standard in the field of interdisciplinary communication studies. It responds to the communication gaps between a range of disciplines in the human and social sciences and humanities and therefore welcomes proposals which integrate a range of diverse approaches (for example, in branches of philosophy, communication theory, social psychology, media studies, and social theory). Particular emphasis will be placed on theoretical innovation, new methodological approaches and the genuinely interdisciplinary work without which communication studies cannot grow.